THE BIRDS OF SÃO TOMÉ & PRÍNCIPE WITH ANNOBÓN

ISLANDS OF THE GULF OF GUINEA

The birds of
SÃO TOMÉ & PRÍNCIPE,
with
ANNOBÓN
islands of the Gulf of Guinea

An annotated Checklist

by

Peter Jones
and
Alan Tye

BOU Checklist No 22

British Ornithologists' Union | British Ornithologists' Club

Department of Zoology, University of Oxford, South Parks Road, Oxford OX1 3PS, UK

www.bou.org.uk | www.ibis.ac.uk | www.boc-online.org

ISBN 0 907446 27 2

British Ornithologists' Union Registered Charity No. 249877

British Ornithologists' Club Registered Charity No. 279583

Photographs: © Phil Atkinson, Martin Dallimer, Peter Jones, Martim Melo, Jorge Palmeirim, Jaime Pérez del Val and Alan Tye

Cover: Dohrn's Thrush-Babbler, Newton's Sunbird and Annobón Paradise-Flycatcher © David Nurney

First Edition 2006

ISBN 0 907446 27 2

Design by Steve Dudley (BOU) and Eng-Li Green (Alcedo Publishing)

Typeset by Alcedo Publishing, Pennsylvania, USA

Printed and bound in China
Phoenix Offset/The Hanway Press

Typeset in Book Antiqua 9pt and Arial 9pt

Dedicated to the memory of
Janet Kear 1933–2004

Contents

Editor's Foreward

Island biota enthral biologists as they provide fascinating glimpses into how evolution works and pose questions about how animals and plants could have reached far flung places in the middle of oceans. Ornithologists in particular are drawn to study islands and have the examples of David Lack's studies of Charles Darwin's Galapagos finches and Robert MacArthur and Edward Wilson's theory of island biogeography to inspire them. The Atlantic islands of the Gulf of Guinea are especially interesting as they form a chain of land masses of increasing distance from the mainland of Africa, linking with a geological formation passing through Mount Cameroon and beyond into the African interior. Thus, the islands' plants and animals can be studied with respect to questions of whether endemism increases with distance from the presumed progenitor populations, rates of colonisation and extinction and, with improvements in molecular techniques, degrees of relatedness.

The Gulf of Guinea islands also have their place in ornithological history. Early explorers and bird collectors, such as Louis Fraser and Boyd Alexander, used Fernando Po (now Bioko in Equatorial Guinea) as a staging post while Karl Weiss, Leonardo Fea, Francisco Newton and José Correia focused on the islands in the late 19th and early 20th centuries. The birds of Bioko are not treated in depth in this checklist, as they have been written about by Jaime Perez del Val in his book *Las Aves be Bioko*, published in 1996. However, Peter Jones and Alan Tye have provided the most complete modern synthesis of the ornithology of São Tomé, Príncipe and Annobón, listed in order of increasing distance from Bioko. The synthesis is based not only on the authors' own observations, made during visits to the islands in the late 1980s, but also on their thorough compilation of the literature and, in addition, it is informed by both authors' personal experiences of studying birds on the African mainland and elsewhere.

Each island has its endemic species and subspecies and, as such, the Gulf of Guinea islands are of global importance for their biodiversity. The highlands of Bioko are part of the designated Cameroon mountains Endemic Bird Area (EBA) and each of the three islands covered by this book is a separate EBA. Of course it is not only birds that are important and, unusually for a British Ornithologists' Union checklist, summaries of endemism in other taxa are given in this volume, supplementing the papers on the flora and fauna of all four of the main islands printed in a special edition of *Biodiversity and Conservation* (1994, volume 3, pages 757-979).

Part of the diversity in the Gulf of Guinea may be attributed to the range of habitats and niches that are found on the islands. The plates illustrate some of the extraordinary scenes and bizarre rock formations that can be witnessed. Yet the lure of wealth from oil and other natural resources is rapidly making the Gulf of Guinea less and less remote and subject to the same pressures that are despoiling so many wildernesses elsewhere. So this checklist, edited until its final stages by the late Janet Kear, is published at an opportune time to draw attention to the unique biodiversity that is found on São Tomé, Príncipe and Annobón and their neighbouring islets and will contribute enormously to the case for its continued conservation. Peter and Alan must be congratulated for bringing this book to fruition and it is a very welcome addition to the British Ornithologists' Union's checklist series.

Robert A. Cheke
BOU Checklist Series Editor
December 2005

Preface

In the first comprehensive analysis of avian biodiversity worldwide, carried out by the then International Council for Bird Preservation (ICBP, now BirdLife International), the three oceanic islands of the Gulf of Guinea emerged as of global importance (Bibby *et al* 1992). On the basis of their high levels of avian endemism, Príncipe, São Tomé and Annobón were included among 218 Endemic Bird Areas (EBAs) worldwide (Stattersfield *et al* 1998). Rather surprisingly for oceanic islands, Príncipe and São Tomé were rated particularly highly, in the top 25% of EBAs, for their species-richness: each supports more than twice the number of restricted-range bird species that would be predicted from their land area alone (Bibby *et al* 1992). More recently, the forests of São Tomé and Príncipe, the island of Annobón and the Tinhosa islets near Príncipe have all been included among the Important Bird Areas (IBAs) of Africa (Fishpool & Evans 2001).

Oceanic islands have long attracted the attention of scientific collectors and evolutionary biologists because of the likelihood of encountering unique forms of life that have evolved in isolation. Until only a few years ago, the Gulf of Guinea archipelago was seldom visited by the scientifically curious, partly because of the islands' geographical isolation and inhospitable climate, but also because, from time to time, they were closed effectively to outsiders for political reasons, often for extended periods. Until recently, therefore, much of our knowledge derived from the few intrepid collectors who had visited the islands at infrequent intervals and rather long ago.

One consequence of this was that, when ICBP came to revise the Red Data Book for birds in the early 1980s, there was no up-to-date information on the status of the birds of the Gulf of Guinea and, frustratingly, the islands were again virtually closed to visitors. Since then, conditions have relaxed considerably, visitors are welcomed once more, and the conservation of biological diversity is high on the political agenda of both the Republic of São Tomé e Príncipe and the Republic of Equatorial Guinea (of which Annobón is part), with generous funds being provided by the European Union's ECOFAC Project for the Conservation and Sustainable Utilisation of Rainforest Resources. It is this re-awakened interest in an archipelago of great biological significance and outstanding natural beauty that has stimulated the present work.

Peter Jones
Institute of Evolutionary Biology
University of Edinburgh
Edinburgh EH9 3JT
Scotland, UK
peter.jones@ed.ac.uk

Alan Tye
Charles Darwin Research Station
Isla Santa Cruz
Galapagos
Ecuador
atye@fcdarwin.org.ec

Acknowledgements

We wish especially to thank Nigel Collar at BirdLife International (then ICBP), who first encouraged us to visit the Gulf of Guinea islands and, with Mike Rands and Paul Goriup, helped to make it possible with funding from the European Commission, ICBP, and IUCN. In São Tomé and Príncipe, we wish to thank Ana Maria de Sá Almeida and Henrique Pinto da Costa for all their assistance during our visits, and John Burlison for his help in the field and subsequently.

We are very grateful to the following people for permission to use their unpublished observations on the islands' birds: Peter Alexander-Marrack, Phil Atkinson, Patrice Christy, Marrack Goulding, Tom Gullick, Peter Kaestner, Sasha d'Assis Lima, Martim Melo, the late Luis Monteiro, and Staffan Reinius.

We are also grateful to the following for providing specialist information and for comments on various parts of the manuscript over a long period: Bill Bourne (seabirds), Bob Cheke (Bioko birds), Gervase Clarence-Smith (history), John Dutton (mammals), John Fa, Estrela Figueiredo (plants), Godfrey Fitton (geology), Mike Harrison (Annobón), Javier Juste B (Annobón birds and place-names), Martim Melo (seabirds, Portuguese and vernacular names), Amberley Moore (ornithological history), Ron Nussbaum (herpetofauna) and Jaime Pérez del Val (birds of Bioko and Annobón).

Many people helped in other ways and we thank: Peter Britton for information on the sources of uncredited data in *Birds of Africa (BOA) Vol. 2*, Patrice Christy for amplifying some of the information contained in his excellent *Guide des Oiseaux de São Tomé et Príncipe*, Godfrey Fitton for the loan of maps and aerial photographs, Hilary Fry for help with the nomenclature and taxonomic sequence used in *BoA*, Angus Gascoigne for help with bibliographical matters, Christoph Hinkelmann for painstaking attempts to trace Hartlaub's specimens in German museums, Adam Jones and Alex Schläpfer for assistance with translations from Dutch and German, the late Stuart Keith and, especially, Linda Macaulay for devoting what must have been ages to a succession of requests for information on specimens and Correia's correspondence and notes in AMNH – their contribution has been considerable. We also thank David Snow for recalling and clarifying events from his 1949 visit to the islands, Rinse Wassenaar (Vogeltrekstation Arnhem) and the late Chris Mead (British Trust for Ornithology) for information on ringed birds, Jaime Pérez del Val for information on specimens in MNCN and help with Spanish references, and Hilary Tye and Sue Wells for help in various ways.

Phil Atkinson, Martin Dallimer, Martim Melo, Jorge Palmeirim and Jaime Pérez del Val very kindly allowed us to include some of their excellent photographs as plates in this checklist; to them, both we and the BOU are extremely grateful.

The unconscionably long gestation period of this work must undoubtedly have tried the patience of the BOU and its previous Checklist Editors, Llew Grimes and the late and much missed Janet Kear, as well as our many correspondents whom we asked repeatedly to update the information they had supplied. Our third Editor, Bob Cheke oversaw eventual publication of the Checklist, whom we thank along with Caroline Dudley (copy-editor) and Steve Dudley (BOU Administrator) who managed the overall production of the Checklist. To all of them we again offer our grateful thanks and hope that the wait has been worthwhile.

Authors' Biographies

Peter Jones began a long association with African ornithology after a DPhil on Great Tits at the Edward Grey Institute, Oxford University, with Chris Perrins. There he met the late Peter Ward, who in 1969 invited him to work in Botswana on the breeding ecology and migration patterns of the notorious bird pest, the Red-billed Quelea. The project was funded by the then UK Government's Centre for Overseas Pest Research (COPR), with whom he worked for 10 years. He spent three of these doing fieldwork around the Okavango Delta and the following three years with Peter Ward in northeastern Nigeria around Lake Chad, still working on quelea ecology but developing an increasing interest in intra-African migration by Palearctic migrants. After more work on birdpests in Ghana and Zambia, he joined Edinburgh University in 1979 as a lecturer in animal ecology. He continued to work on intra-African migration by Palearctic and Afrotropical species and had the chance for two visits to São Tomé and Príncipe on behalf of ICBP (now BirdLife International) and IUCN, first with Alan Tye in 1987 and then with John Burlison in 1988. Together with his students, Peter has maintained a strong interest in the evolution and phylogeography of the Gulf of Guinea avifauna ever since. Subsequent unique diversions included several months studying the ecology of the endemic birds of Henderson Island in the Central Pacific with the Sir Peter Scott Commemorative Expedition to the Pitcairn Islands 1991-92, and 5 months in 2003 with the Swiss Ornithological Institute's Sahara Project counting trans-Saharan migrants stopping over in the oasis of Ouadâne, Mauritania. At present he is collaborating with Bob Cheke, the current Checklist Editor, to forecast the migrations of queleas in southern Africa using rainfall data derived from satellite imagery. Peter was Editor of *Ibis* 1988-1993 (initially with Janet Kear) and BOU Vice-President 1993-1997; he also edited the European Ornithologists' Union's journal *Avian Science* from 2001 to 2003.

Alan Tye was born in Nottingham and attended school there. After a degree in Natural Sciences and PhD on Wheatears and Fieldfares at Cambridge University and Monks Wood Experimental Station, a lifelong interest in wildlife and a fascination with faraway warm places led to a determination to work in the tropics. His first major job was as Lecturer in Zoology at the University of Sierra Leone, where began a long association with West African ornithology. After five years in Sierra Leone, he returned to the UK, working for the then Nature Conservancy Council, plus bird surveys in Kuwait and S.E. Asia, and then, in 1987, came an unexpected invitation to take part in an exploratory survey of endemic birds and bird habitats in Sao Tomé and Príncipe. It was while on the islands in 1987 that Peter and Alan came up with the idea of this checklist. Alan began working on the book in the evenings while on his next job, based in Amacayacu National Park in the Colombian Amazon, where he led the BOU's two-year expedition and worked on frugivorous birds and their food plants. After returning from Colombia in 1989, Alan worked on a Houbara Bustard project in Saudi Arabia, and re-established a West African bird link by taking on the editorship of *Malimbus*. A post with IUCN followed, as Forest Conservation Advisor on their East Usambara Conservation and Development Project in Tanzania, where the only research that was part of the job was on invasive plants. That experience led to an appointment in 1996 as Head of the Botany Department at the Charles Darwin Research Station in Galapagos. Alan's research is, of course, principally tropical, with island biology, conservation science and invasive species management as special interests. He continues to edit *Malimbus* and in 2005 added the journal *Galapagos Research* to his editing responsibilities. He chairs the IUCN Galapagos Plant Specialist Group, which he created, is a member of the Invasive Species Specialist Group, and continues to think about West African birds while watching Darwin's finches.

List of tables

List of figures

List of plates

INTRODUCTION

Location and Geographical Limits

The four islands that bisect the Gulf of Guinea (Fig. 1a) are, from north to south:

*   Bioko: (3°12′N–3°47′N, 8°25′E–8°56′E) formerly the Spanish possession of Fernando Po, now part of the Republic of Equatorial Guinea. Bioko is the largest of the four islands, some 2027 km^2 in area (roughly 35 km x 72 km) and the closest to the mainland, lying on the continental shelf only 32 km from the coast of Cameroon.

*   Príncipe: (1°32′N–1°43′N, 7°20′E–7°28′E) formerly Portuguese and now part of the Democratic Republic of São Tomé and Príncipe. It lies 210 km SSW of Bioko and 220 km W of the African mainland. Its total area is 139 km^2 (c 17 km x 8 km).

*   São Tomé: (0°25′N–0°01′S, 6°28′E–6°45′E) formerly Portuguese and now part of the Democratic Republic of São Tomé and Príncipe. It lies 150 km SSW of Príncipe and 255 km W of Gabon, and the equator passes through the Ilhéu das Rolas at its southern tip. Its total area is 857 km^2 (47 km x 28 km).

*   Annobón: (1°24′S–1°28′S, 5°36′E–5°38′E) now a province of the Republic of Equatorial Guinea. It is the smallest and most remote, lying 180 km further to the SSW of São Tomé and 340 km from the mainland. Its total area is 17 km^2 (6 km x 3 km).

Bioko lies in shallow seas only 60 m deep. During the last glaciation, sea levels were lowered sufficiently to connect Bioko to the African mainland by a broad landbridge (Fig. 1b). Rising sea levels would have isolated Bioko once more only c 11,000 years ago (Eisentraut 1965, Moreau 1966, Lambeck & Chappell 2001). Its flora and fauna have not therefore evolved in oceanic isolation as has been the case on the outer three islands of Príncipe, São Tomé and Annobón, which are separated by seas over 1800 m deep and, clearly, have never been connected by land to one another nor

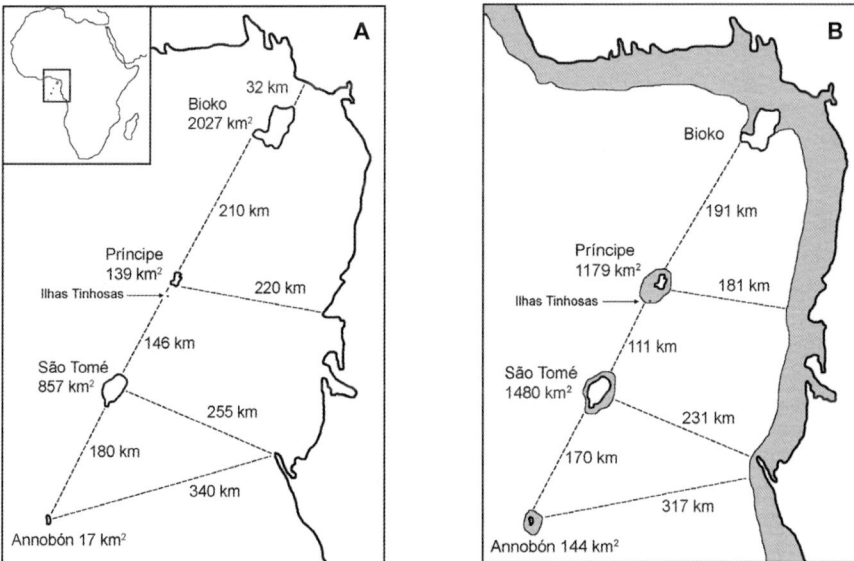

Figure 1 Map of the Gulf of Guinea showing island areas, and distances between them and the African coast (A) at the present day and (B) during the last glacial maximum (c 25,000–19,000 BP)

with the African mainland. Bioko's recent avifaunal history, therefore, more closely parallels that of coastal Cameroon and also, because of the island's high elevation (up to 2850 m), that of Mt Cameroon and the Bamenda Highlands. This is reflected in the fact that, while Bioko has about 140 breeding bird species, only one is endemic, the Fernando Po Speirops *Speirops brunneus*, although at least 46 others (33% of the resident avifauna) are subspecifically distinct from neighbouring mainland forms (Pérez del Val *et al* 1994, Pérez del Val 1996). For these reasons, and because it harbours many more species that are not found on the other islands, we have excluded Bioko from the present Checklist and restricted our survey to the three oceanic islands with their relatively depauperate avifaunas but high levels of species endemism. For completeness, however, a checklist of Bioko's avifauna is included in Appendix 1.

History

São Tomé and Príncipe were discovered by two Portuguese navigators, João de Santarém and Pedro de Escobar, in late 1470 and early 1471 (Henriques 1917 in Exell 1944). From the names given to the islands it is probable that São Tomé was sighted on 26 December 1470 (St Thomas's Day) and Príncipe (originally São Antão) on 17 January 1471 (St Anthony's Day). It is also generally accepted that Annobón (originally Ano Bom = New Year's Day) was discovered by the same expedition on 1 January 1471 but Valentim Fernandes (1506–10, reprinted and translated as Monod *et al* 1951) gave the date of discovery as 1 January 1501 by a Portuguese from São Tomé, Fernão de Mello. All three islands were said to be uninhabited when first discovered. Unless there had been previous peoples who disappeared without trace, the islands had remained completely unmodified by man until this time. Portuguese jurisdiction over São Tomé was established in 1485 when João de Paiva took the first colonists there, but further settlement did not occur until 1493. The settlement of Príncipe began in 1500, when the Portuguese Crown gave the island to António Carneiro.

Soon after settlement, the cultivation of sugar cane began in the northern parts of both São Tomé and Príncipe. By 1550, sugar had become an important export and production reached a peak of 12,000 tonnes in 1578. The lowland forest over much of São Tomé had therefore been destroyed by this date, except in the Angolares region of southeastern São Tomé. This was occupied by Africans from Angola, said to have escaped from a wrecked slave-ship in 1540, who had established a semi-independent colony. They continued to cultivate the Angolares region for the following three centuries by means that apparently did not cause much forest destruction. Forest in the north of Príncipe was also partly cleared at this time but survived into the 20th century on land that was too poor for agriculture.

Sugar cane production declined during the 17th century and the islands subsisted by producing food for slave traders in transit. Further change did not occur until coffee was introduced from Brazil to São Tomé in 1800 and to Príncipe in 1802. Coffee estates (*roças*) were established on São Tomé up to 1200 m asl, causing more forest destruction. Cocoa was brought to Príncipe from Brazil in 1824 and introduced from Príncipe to São Tomé in 1855. Coffee production declined after 1870 because of pathogens and exhaustion of the soil, and much of the old coffee area was replanted with cocoa. The Angolares region of São Tomé was taken over finally and planted in 1875. Between 1908 and 1919, São Tomé (with Príncipe) was the world's largest cocoa producer, exporting over 30,000 tonnes annually. Cocoa production declined after this and some plantations were abandoned, but this decline was most marked at independence from Portugal in 1975.

According to Valentim Fernandes (Monod *et al* 1951), Annobón was settled in 1503. In 1778 it was ceded by Portugal to Spain, which then abandoned it and its population until a Catholic mission was founded in 1865; Spanish authority was re-established in 1885. Because of its small size and geographical isolation, Annobón seems only ever to have had a subsistence economy based on agriculture, fishing and some whaling. It became independent from Spain in 1968 as a province (Pagalú) of Equatorial Guinea.

The History of Avifaunal Investigation

Well-referenced accounts of the early history of ornithological investigations in the islands were given by Tommaso Salvadori (1903a,b,c), while further details of all these and later visits have been provided by Réné de Naurois in various papers. This brief history draws heavily on these, with more recent references where necessary (eg Fry 1961 for Annobón). We have tried to rectify numerous mismatches and contradictions concerning localities and dates wherever possible.

Príncipe

The earliest known endemic bird from Príncipe, the Príncipe Glossy Starling *Lamprotornis ornatus*, was described in 1800 by Daudin from Levaillant's specimens, but the island's avifauna did not become well known until the 1840s. Although the British Government's Expedition to the River Niger used the Gulf of Guinea islands as a base, collecting several new species on São Tomé and Annobón (see below), they visited Príncipe only briefly, in October–December 1841 and June–July 1842, and no specimens were recorded as having been collected there (Allen & Thomson 1848, A. Moore pers comm). A few years later, however, Carl Weiss collected the Príncipe Drongo *Dicrurus modestus* (described by Hartlaub 1849) and the Príncipe Golden Weaver *Ploceus princeps* (described by Bonaparte in 1850 from a specimen sent to Paris). Confusion surrounded the provenance of other Príncipe specimens from this period, resulting in Hartlaub (1857) wrongly describing the Príncipe Speirops *Speirops leucophaeus* as coming from Gabon and the Príncipe (Hartlaub's) Sunbird *Anabathmis hartlaubii* from Angola (Hartlaub 1857), while Gray (1862) described the nominate race of the Príncipe Seedeater *Serinus rufobrunneus* without any locality.

These confusions were eventually sorted out with the help of the most important early collection from Príncipe, that of H. Dohrn and J.G. Keulemans, who spent six months on the island from April to September in 1865 (Dohrn 1866, Keulemans 1866). This collection is now in the museum at Szczecin. Dohrn and Keulemans not only redescribed all of the above species but also found Dohrn's Thrush-babbler *Horizorhinus dohrni*, a curiously late discovery given the bird's commonness and its loud and attractive song. They also obtained specimens of the extremely rare endemic subspecies (*rothschildi*) of the Olive Ibis *Bostrychia olivacea*. Although Dohrn was a meticulous worker, he seems not to have been a good field observer, while Keulemans, if sometimes over-exuberant, was a better naturalist and his field notes remain a valuable source of information.

The next important period of bird collecting occurred during the decade or so from 1885 to 1895, when Francisco Newton assembled a comprehensive collection of birds and other terrestrial vertebrates from Príncipe, São Tomé and Annobón for the Lisbon Museum, which were described in a long series of papers by Bocage. Like Keulemans, Newton was a keen observer and made valuable field notes on the birds' behaviour and ecology. Sadly, most of these are now lost. Bocage published a little of Newton's ancillary information but it seems that although Newton 'held some kind of official position as Government Naturalist. . . he lived for a time in a tent on some waste ground in the town of São Tomé in such poverty that he could scarcely obtain paper to write tickets for his specimens' (Exell 1944). The letters he did write to Bocage to accompany his specimens, along with almost all the specimens themselves, were destroyed in the disastrous fire in the Museu Bocage in Lisbon in 1978.

Newton was followed by Leonardo Fea, collecting for the Genoa museum in January–June 1901. Fea obtained a single specimen of the ibis, which was not seen by Newton, and also discovered the rare Príncipe race *xanthorhynchus* of the Gulf of Guinea Thrush *Turdus olivaceofuscus*.

Two important collections were made in the early decades of the 20th century. Boyd Alexander visited the island in February 1909 collecting for the British Museum (Bannerman 1914) and José Correia and his wife Virginia collected there in August–October 1928 for the American Museum of Natural History, New York (Fig. 2; Amadon 1953). Like Newton and Keulemans before them, Alexander and Correia both kept extensive field notes on the birds they collected. Some of Alexander's notes were retrieved after his tragic murder on 2 April 1910, near Abéché in present-day Chad, and were quoted extensively by Bannerman (1914, 1915a,b) when he later described Alexander's collections from Príncipe, São Tomé and Annobón. Alexander's field notebook is now held at Tring. Correia's valuable and entertaining diary (1928–29b) remains unpublished but his collection was described by Amadon (1953) in the most important synthesis to date on the avian zoogeography of the Gulf of Guinea islands.

David Snow spent two weeks on Príncipe in 1949 (Snow 1950), providing extensive notes on birds in the field. He was followed by Frade in 1954 (Frade 1958, Frade & Santos 1977) and later still by Naurois between 1963 and 1973, who produced over 20 papers on the biogeography and taxonomic affinities of the islands' birds (eg Naurois 1983a, see Systematic List). The significant findings of these visits were the addition of *Zoonavena thomensis* to the list of breeding birds, the description of an endemic subspecies of the Príncipe Seedeater, *Serinus rufobrunneus fradei*, from the tiny Ilhéu Caroço (= Ilhéu Boné de Jóquei; Plate 28) off the southeastern end of Príncipe, and the first indication of the occurrence of an unknown scops owl on Príncipe (Naurois 1975a). After Naurois's last visit there were no more ornithological surveys of Príncipe until the ICBP surveys in 1987 and 1988 (Burlison & Jones 1988, Jones & Tye 1988) and the 1990 University of East Anglia expedition (Atkinson *et al* 1991). None of these found any evidence of the continued existence of either the thrush or the ibis but, happily, the ibis was eventually rediscovered in 1991 (Sargeant 1994) and the thrush in 1997 (S. Lima, J Rosseel *in litt*). Significant research on the seabirds of Príncipe and the Ilhas Tinhosas has been carried out by a team from the University of the Azores in 1996–97 (Monteiro *et al* 1997). More recent studies of the

Figure 2 José Correia (1881–1954), who collected on the Gulf of Guinea islands in 1928–29 (photo by Rollo Beck on the AMNH South Seas Expedition 1923–26; Beck Collection, California Academy of Sciences)

landbirds, by Jonathan Baillie and Angus Gascoigne in 1999 and Martin Melo and Rita Covas at present, are in preparation.

São Tomé

The earliest record of any endemic species of São Tomé seems to be that of the São Tomé Green Pigeon *Treron sanctithomae*, first mentioned by Gustav Marcgrav in 1648 and described by Gmelin in 1783. However, the avifauna remained poorly known until almost the middle of the 19th century and, even then, the original descriptions were beset with confusion about the provenance and true identity of the specimens. Several of the earliest species to be described came from collections made in 1841 while three Royal Navy ships (*Wilberforce*, *Albert* and *Soudan*), supporting the British Government's Expedition to the River Niger, were using the Gulf of Guinea islands as a base (Allen & Thomson 1848). William Allen commanded the *Wilberforce*, Thomas Thomson was the Assistant Surgeon and Louis Fraser was the Expedition's Naturalist. The *Wilberforce* sailed from Bioko to Príncipe, São Tomé, the islet of Rolas, Annobón and Ascension Island during October–November 1841, returning to Príncipe and Bioko in June 1842. The *Albert* was at Rolas and Príncipe in December 1841 en route to Ascension Island, and at Príncipe again in June and July 1842 (A. Moore pers comm). Considerable confusion surrounded their specimens, engendered in part by the repeated loading and unloading of cargoes between ships and between islands. Thus, the provenance of the endemic São Tomé Paradise Flycatcher *Terpsiphone atrochalybeia*, first described by Thomson (1842), was wrongly given as Fernando Po (Bioko). While Fraser first described the endemic São Tomé Green Pigeon from Rolas (it does not occur on the islet nowadays), several mainland species recorded by Allen and Thomson (1848) from Rolas have never been seen on the islands (Salvadori 1903b). Other misidentifications resulted simply because African bird taxonomy was then in its infancy. The Giant Weaver *Ploceus grandis*, described from the same expedition by Fraser as *P. collaris*, was only properly recognised by Gray in 1844. The Maroon Pigeon *Columba thomensis*, identified as *C. trigonigera* and recorded on the islet of Rolas by Allen and Thomson (1848), was not recognised as specifically distinct until 1896 by J.G. Barboza do Bocage.

The most important early collection, however, was made between 1847 and 1850 for the museums of Hamburg and Bremen by Carl Weiss. Weiss, a doctor, arrived in São Tomé on 20 August 1847, then went to Ghana and Príncipe before returning to São Tomé. Most of his specimens were described in Hamburg by Gustav Hartlaub (summarised in Hartlaub 1850, 1857). The 26 species sent back included the endemic São Tomé Scops Owl *Otus hartlaubi* (but recognised as specifically distinct only by Giebel (1872)), the Gulf of Guinea Thrush *Turdus olivaceofuscus*, the São Tomé Weaver *Ploceus sanctithomae*, the São Tomé Speirops *Speirops lugubris* and the São Tomé Oriole *Oriolus crassirostris*, as well as endemic races of some mainland species.

During this period, the colonial power, Portugal, was also amassing its own collections from São Tomé. By 1867 most specimens from São Tomé at the Lisbon Museum were listed as being either from 'Dr Nunes' or from the collection of King Pedro V, given to the museum by Luiz I in 1863 (Bocage 1867) but none of these broke new ground. A decade later, in 1879 and 1880, R Greeff carried out extensive research on other terrestrial vertebrates but did not concern himself with the avifauna.

The next important collections on São Tomé were made between 1885 and 1895 by Francisco Newton for the Lisbon Museum. Newton's collections (summarised in Bocage 1904) provided several new endemic species: São Tomé (Newton's) Fiscal *Lanius newtoni* (first collected in 1891), Giant Sunbird *Dreptes thomensis* (1889), São Tomé (Newton's) Sunbird *Anabathmis newtonii* (1887) and the São Tomé Grosbeak *Neospiza concolor* (1888). Newton also sent Bocage's Longbill (São Tomé Short-tail) *Amaurocichla bocagii* to Bocage, but Bocage in turn sent it to R.B. Sharpe in London, who described it in 1892 (Sharpe 1892a). Newton's specimens of the São Tomé Spinetail *Zoonavena thomensis* were later described as specifically distinct by Hartert (1900), and his Dwarf Ibis *Bostrychia bocagei* by Chapin (1923), both of which Bocage had originally described as island subspecies of mainland forms. Newton's collections enabled Bocage to describe the Maroon Pigeon as an endemic species *Columba thomensis*, the São Tomé specimens of the seedeater *Serinus rufobrunneus*, already known from Príncipe, as an endemic race *thomensis*, as well as several island subspecies of mainland birds.

During the same period, A.F. Moller, who was on São Tomé between 23 May and 25 September 1885, obtained some birds while collecting plants for the Coimbra Botanic Gardens (Lopes Vieira 1887). It was from his specimens that the São Tomé Prinia *Prinia molleri* was named by Bocage (1887a), though it was perhaps this species that had earlier been referred by Hartlaub to *Drymoica ruficapilla* (see Systematic List).

By the turn of the century, no more endemic land-bird species remained to be discovered. However, an important collection made for the museum at Genoa by Leonardo Fea between 1899 and 1901, and described by Salvadori (1903b), included the hitherto undescribed São Tomé race *feae* of the white-eye *Zosterops ficedulinus*, already known from Príncipe.

Boyd Alexander visited São Tomé in January–February 1909 collecting for the British Museum (Bannerman 1915a), and the Correias spent two periods there between May 1928 and March 1929 (Correia 1928–29b). Their collections are valuable not only for their specimens, especially the Correias', but also for the indications that their field notes give us of the distribution and status of the endemic species at a time when the agricultural exploitation of São Tomé was at its peak. Some species were evidently very rare by the end of the 19th century. Neither Fea nor Alexander saw the ibis, fiscal, longbill or grosbeak, which were then known only from Newton's collections, yet the Correias managed to rediscover all except the last.

Twenty years then passed until the next ornithological survey of São Tomé, in 1949, when an Oxford University team spent September there and published the first detailed field notes for most of the endemic birds (Snow 1950). There followed a Portuguese scientific mission, lasting three months in 1954, to collect birds and to assess the status of those that might require protection under Portuguese law (Frade 1958). After another long interval, R. de Naurois visited the islands several times in 1963 and 1970–73, the result of which was an extensive series of papers dealing with all that was then known of the ecology and systematics of all of the endemic and many of the other indigenous species (see References). Naurois also transcribed some of Francisco Newton's papers before they were destroyed. Naurois's work culminated in the publication of the first book to deal exclusively with the birds of all three oceanic Gulf of Guinea islands (Naurois 1994). During the same period, the archipelago's seabirds first received some limited attention from the University of Miami's R/V *Pillsbury* expedition in 1964–65 (Robins 1966).

A further decade passed after de Naurois's last visit before ornithologists next visited São Tomé. A team from the Dresden Museum visited São Tomé for three weeks in June 1983 (Günther & Feiler 1985), while a doctor resident for two years in the early 1980s kept useful bird notes (Reinius 1985). Interest in the Gulf of Guinea islands was then given a considerable boost by the publication by the International Council for Bird Preservation (ICBP, now BirdLife International) of the third edition of the bird Red Data Book (Collar & Stuart 1985), which considered seven endemic species of the islands to be threatened. However, the lack of up-to-date information about the islands was highlighted by the fact that four of these species, the Dwarf Ibis, Bocage's Longbill (São Tomé Short-tail), São Tomé (Newton's) Fiscal and the São Tomé Grosbeak had not been seen for over fifty years, because of a lack of adequate surveys in the more remote parts of São Tomé. In April 1987, Stephen Eccles spent a week on São Tomé and reported the rediscovery of Bocage's Longbill

(Eccles 1988). In 1987 and 1988, ICBP initiated three European Community-funded visits to survey the birds and to make recommendations for their future conservation (Burlison & Jones 1988, Jones & Tye 1988, Harrison & Steele 1989, Jones *et al* 1991, 1992). There followed a University of East Anglia expedition in 1990 (Atkinson *et al* 1994) that was remarkable for rediscovering two more of the endemic species, the Dwarf Ibis and São Tomé (Newton's) Fiscal, which had long been feared extinct (Collar & Stuart 1985). The remaining enigma, the São Tomé Grosbeak, was finally rediscovered in 1991 (Sargeant *et al* 1992), more than a century after it had last been seen. A multi-disciplinary expedition from the Dresden Museum visited São Tomé in March–April 1991 and added further to our knowledge of the birds (Nadler 1993). The islands' seabirds finally received more detailed study from the University of the Azores seabird expedition in 1996–97 (Monteiro *et al* 1997).

São Tomé and Príncipe are now on many birders' itineraries, while further ornithological survey work is being carried out as part of the 'ECOFAC' project, a seven-country regional programme funded by the European Union for the sustainable use of rainforest resources in Central Africa (Anon 1994). Among the outputs of this programme was the first field guide to the birds of São Tomé and Príncipe (Christy & Clarke 1998).

Annobón
Annobón was visited by the Niger Expedition from 27 October to 5 November 1841 and again in late December 1841. Although Louis Fraser collected the Annobón Paradise Flycatcher *Terpsiphone* (*rufiventer*) *smithii* during this period, it was not until 50 years later, when Francisco Newton spent a little over a month there from 19 November 1892 to early January 1893, that the avifauna was first seriously collected (Bocage 1893a). Newton sent 14 species back to Bocage in Lisbon, including the island's second endemic species, the Annobón White-eye *Zosterops griseovirescens*, described by Bocage (1893a). Leonardo Fea visited Annobón in April and May 1902 and caught eight species, including endemic subspecies of Scops Owl *Otus scops feae* and Lemon Dove *Columba larvata hypoleuca*, both described by Salvadori (1903c), neither of which had been collected by Newton. Boyd Alexander collected on the island for a week in February 1909 (Bannerman 1915b) but found nothing new. After a one-day visit by W.P. Lowe in December 1910, there was a long gap until Aurélio Basilio visited between August and November 1955 (Basilio 1957), followed by Hilary Fry in July and August 1959 (Fry 1961). There then followed another long gap until interest in the island revived once more (Harrison 1990, Pérez del Val 2001).

Geology and Geomorphology
The islands of the Gulf of Guinea form the southern part of the Cameroon Line of Tertiary to Recent volcanoes, which stretches for 1600 km from the outermost island of Annobón, crossing the continental shelf at Bioko and including Mts Cameroon, Manengouba and Oku, the Biu and Ngaoundéré Plateaux and the Mandara Mountains on the African mainland to the north. There is no evidence of current volcanic activity on the islands and only Mt Cameroon remains active on the continent.

Príncipe
Príncipe (Fig. 3) comprises two distinct regions: a relatively flat, low-lying basalt platform in the north with small hillocks below 180 m, contrasting sharply with the mountainous central and southern region (Plates 4, 5 & 23). The main peaks, Pico (948 m), Mencorne (935 m) and Carriote (830 m) form a topographic barrier between the two parts of the island. A large isolated plug slightly farther north, Pico Papagaio, rises to 680 m. The island is surrounded by a broad submarine shelf that was exposed during the last ice age when sea-levels were lowered.

The oldest exposed rocks are breccia and basalt derived from a much earlier submarine phase in the evolution of the island and contain blocks of olivine tholeiite dated at 31 million years (Ma). On top of this are two sub-aerial series of basaltic flows. The oldest, exposed in the north of the island, are 24 to 19 million years old (Ma) and the youngest date from 5.6 to 3.5 Ma. The large number of phonolite, trachyte and tristanite plugs in the south of the island date from 6.9 to 3.5 Ma (Fitton & Hughes 1977, Fitton 1987, Lee *et al* 1994).

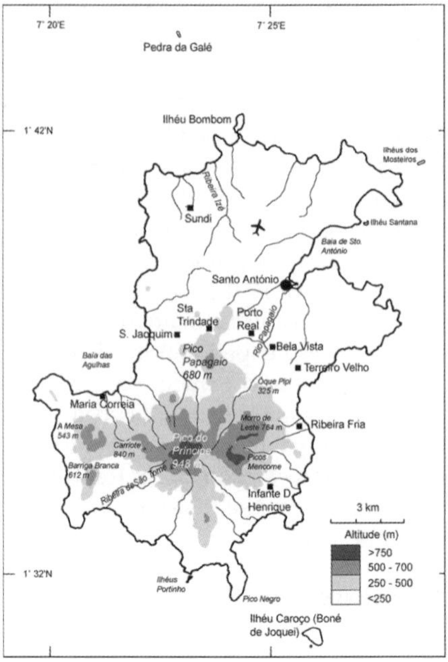

São Tomé

São Tomé (Fig. 4) rises steeply from about 3000 m below sea-level to over 2000 m asl in its centre. According to Mitchell-Thomé (1970), the northeast part of São Tomé is a bevelled platform of basalt sloping gently seawards from elevations of 200–300 m. The rivers run in slightly incised valleys appearing as bouldery and sandy trenches. Some run into marshy areas, often brackish with mangroves, separated from the sea by banks of white calcareous sand or pebbles and cobbles, as at Morro Peixe in the north and Malanza in the south (Plates 19 & 22). Evidence of recent uplift is shown by raised beaches along the coast and fluvial terraces in several valleys.

In the central, northeastern and southern areas, intermediate in elevation between the coast and the western mountains, there are smaller level areas of basalt, alternating with narrow, deep, steep-sided valleys amid a sugarloaf landscape of volcanic cones and plugs, of which Cão Grande (663 m) is the most spectacular (Plates 1–3). The mountainous centre and southwest of the island is a deeply dissected volcanic landscape of phonolite and trachyte spines with impressive waterfalls, steep slopes and plentiful talus deposits, falling very steeply to the sea in the west and rising along a sharp ridge to 2024 m at the Pico de São Tomé in the north.

The oldest rocks are conglomerates, sandstones and shales accompanied by oil seepages. They have not been dated with certainty but are possibly Cretaceous (J. Hedberg in Fitton 1987). The oldest dated rocks are quartz trachytes from 15.7 Ma, and the plug forming the Ilhéu das Cabras has been dated at 13.0 Ma; all other dated volcanic rocks are much younger at 0.1 to 7.6 Ma (Fitton & Dunlop 1985).

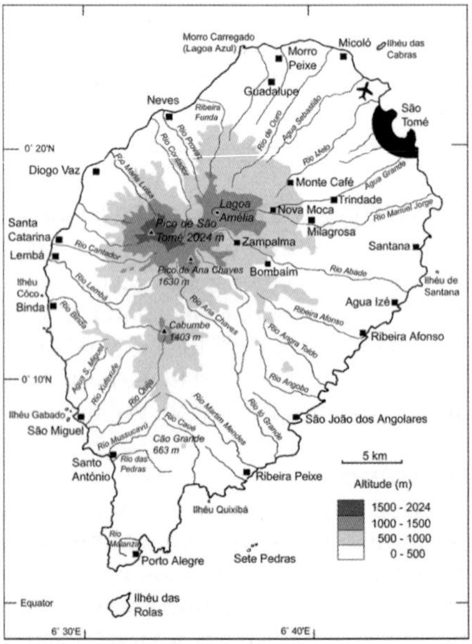

Figure 3   Map of Príncipe

Figure 4   Map of São Tomé

Annobón

Annobón (Fig. 4, Plates 29 & 30) rises to 700 m at Santamina in the south, with the crater of Quioveo (640 m) comprising the central part of the island. Pico do Fogo (450 m), a trachyte plug, rises in the northern rim of the crater and just below this is a small crater lake, Lago A Pot, 220 m asl. Of the many valleys, only three contain permanent streams, including that which drains the crater lake; in prolonged dry periods even the lake itself has been known to dry up.

The oldest rocks, exposed around the coast, are breccias intruded by basaltic dykes but neither of these formations has been dated. These are overlain by two series of basic lavas dated at 4.8 Ma and 0.2 to 1.0 Ma (Lee *et al* 1994), much younger than formerly estimated (see Fitton 1987).

The age of the Cameroon line

The Cameroon line has been active in its continental sector since the Palaeocene (65 million years ago) and along its

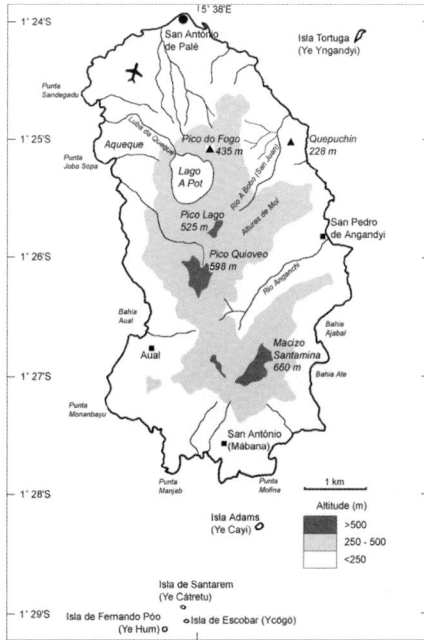

Figure 5 Map of Annobón

whole length at least since the Oligocene (>30 million years ago). There is no systematic progression in the ages of volcanism for the Cameroon line as a whole and activity has continued over its whole length in the past million years, which appears to rule out its origin as a 'hot-spot trail' (Fitton & Dunlop 1985). Nevertheless, recent data suggest that the Cameroon line has been expanding gradually oceanwards and might originate from a sublithospheric 'hot zone' fed periodically by deep mantle plumes; the earliest erupted magmas were in the continental sector (>35 million years ago), while the three oceanic islands have had episodic and protracted volcanic histories, beginning at least 31 million years ago on Príncipe, >13 million years ago on São Tomé and <5 million years ago on Annobón (Lee *et al* 1994). Like all volcanic islands, those of the Gulf of Guinea are likely to have supported life continuously throughout their time as dry land, though it is impossible to say which elements of the biota would have survived or perished during particular periods of volcanism.

Climate

The climates of the three islands are similar (data for Annobón in Fry 1961, Loboch 1962; for São Tomé and Príncipe in Bredero *et al* 1977). The islands' high relief intercepts the prevailing moist southwesterly winds throughout the year, so that rainfall in their southwestern parts probably exceeds 7000 mm annually on São Tomé and 5000 mm on Príncipe. Annobón is probably less wet because of its much smaller size and lower relief; the only rainfall data available (*c* 1000 mm per year) were recorded at the north end of the island.

São Tomé and Príncipe have a rainy season from September to May and their driest period, known as the *gravana*, in July–August (Figs 6 & 7). There is usually also a dry period of less importance in December and early January known as the *gravanito*. Annobón, south of the equator, has a more or less continuous rainy season, the *tendaua*, from October to April (but with rather less rain in January and February) and a single dry season, the *jalma*, from mid May to the end of October.

(a)   (a)

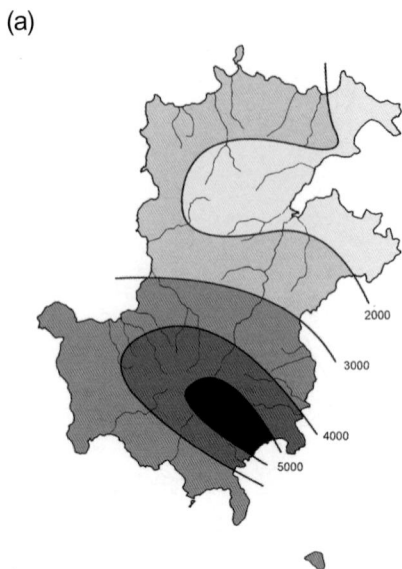

(b)   (b)

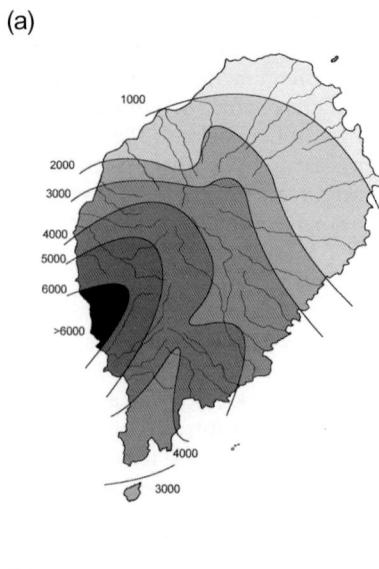

Figure 6 (a)   Mean annual isohyets (mm) and (b) mean monthly rainfall on Príncipe (BDPA 1985)

Figure 7 (a)   Mean annual isohyets and (b) mean monthly rainfall in NE (dark shading) and SE (light shading) São Tomé (BDPA 1985)

The northern areas of each island receive much less rainfall, usually around 2000 mm, at low altitude because of the rain-shadow effect of the high ground to the south; in the extreme north of São Tomé it may be as little as 600 mm per year. It is in the northern areas of each island that the dry season is most marked. The southwestern parts and high interiors of São Tomé and Príncipe are wet almost throughout the year; on São Tomé it may rain for a few hours each day in the afternoon or at night, even during the dry periods (J.-J. Bolyn pers comm 1987).

At sea-level it is hot, with daily maximum temperatures of 22–33°C and a mean humidity of 80%. At higher altitude, around 700 m asl, the mean maximum temperature of about 25°C is similar to that at the coast but the absolute minimum is considerably less, around 9°C. It is much colder at the summits at night, though frosts are unknown. Winds on the islands are generally light, but strong winds accompany the change of seasons, at least on Annobón in March–April and in October. On São Tomé, violent electrical storms are common at high altitudes. Because of the permanent thick cloud in the centres and southern parts of the islands right to the summits, light levels are very low. Data for sunshine hours are few but values from Monte Café at 700 m on São Tomé are probably typical of the northern slopes of the mountains, with daily averages ranging from 1.5 h in August and September to 3.5 h in May. In the southwest of each island the sunshine hours are probably less than this but in the north will be much more, especially in the dry seasons.

Vegetation

The fullest account of the vegetation of Príncipe, São Tomé and Annobón was given by Exell (1944, 1956, 1963, 1973), who collected extensively on the islands in 1932–33. Additional information and floristic analyses were given by Monod (1960), who visited São Tomé and Príncipe briefly in 1956, while the afro-montane elements in the flora of São Tomé have been discussed by White (1983, 1983–84). Publication of an authoritative 'Flora' (Liberato & Espírito Santo 1972–82) is currently in abeyance. Exell's work remains an excellent source of botanical information but is in the process of much-needed revision by Figueiredo (1994a,b,c, 1995, 1998), who has also produced a botanical bibliography (Figueiredo 1994d).

Príncipe

Príncipe was once completely covered in forest and seemed to lack the fire-climax grassland and dry woodland at its northern end, where it might be expected by analogy with São Tomé (see below). Most of the accessible regions of the island were cleared and planted long ago with cocoa and coffee, with coconuts and bananas in some areas. Almost all of the remaining primary forest was said to have been destroyed during a campaign to eradicate sleeping sickness in 1906 (Exell 1944), but reports held in the government archives suggest that the effort required to achieve this was largely beyond the capabilities of the control teams. Much of the southern part of the island must have been left untouched (G. Clarence-Smith pers comm), though the forest surviving on poor soils at the north of the island was probably lost at this time (A. Gascoigne pers comm). The surviving lowland forest (Fig. 8, Plate 6) resembles the 'Lower Rainforest Region' of São Tomé (see below), though

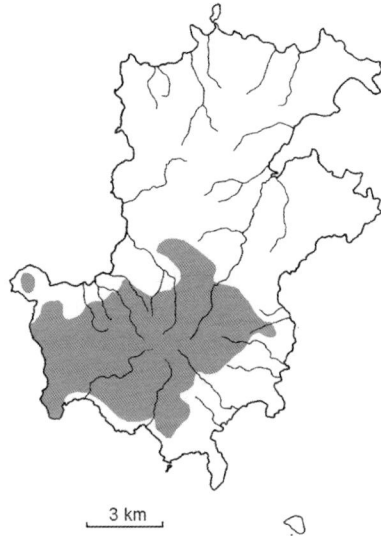

Figure 8 Approximate distribution of surviving primary forest on Príncipe (adapted from Christy & Clarke 1998)

relatively impoverished. Towards the summit of the Pico do Príncipe the forest assumes a slightly more montane character, similar to the transition zone between lowland and montane forest on São Tomé, but the altitude is too low to support mist-forest vegetation. Endemic trees include *Rinorea insularis*, *Ouratea nutans*, *Casearia mannii*, *Croton stelluliferus* and *Erythrococca columnaris*. Príncipe is particularly rich in Euphorbiaceae, including five endemic species.

São Tomé

Apart from some very small areas of mangrove and sand dune on the coast, Exell considered the original vegetation of São Tomé to have been rainforest (known locally as *obô*) uniformly covering the island from sea-level almost to the summit of the Pico (Plate 11). It would seem more likely, however, that near the coast in the north of the island where rainfall is low, the original vegetation would have been a dry woodland. The few small scattered patches of dry woodland remaining in gullies appear natural and may be the last remnants of this vegetation type. The relatively large areas of grassland that remain uncultivated, with scattered acacias and baobabs *Adansonia digitata*, are frequently burned in the dry season (Plates 15–18). It is unknown how much of this grassland formation might be a natural vegetation type; more likely it indicates the former extent of sugar cane cultivation that later became pasture for cattle grazing (A. Gascoigne pers comm).

In the northern half of the island, the land has been brought almost entirely under cultivation from sea-level to about 800 m asl. At present it is mostly given over to cocoa plantation with smaller areas of coffee, bananas and coconut. The 800-m contour marks the upper limit at which cocoa can be grown successfully in São Tomé because of the adverse climate and the prevalence of fungal disease (Lains e Silva 1958); most is now grown below 600 m. Where cocoa and coffee have remained in production under shade trees the result is a 'shade forest' (*floresta de sombra*) of tall trees above an understorey of cocoa or coffee bushes, whose relatively complex structure resembles that of the original rainforest (Plates 12–14).

The lowland rainforest in the wet southern half of São Tomé was largely replaced with coconuts near the coast and cocoa farther inland. However, areas of untouched lowland forest remain in the catchments of the major rivers flowing to the southeast and southwest. The abandonment of so many of the plantations has enabled a dense growth of secondary forest (*capoeira*) to reclaim extensive areas at low altitudes in the south, so that it is now sometimes difficult to tell where native forest and abandoned cultivation begin and end.

Within the original rainforest, Exell distinguished three zones (Fig. 9; see Exell 1944 and Figueiredo 1995):

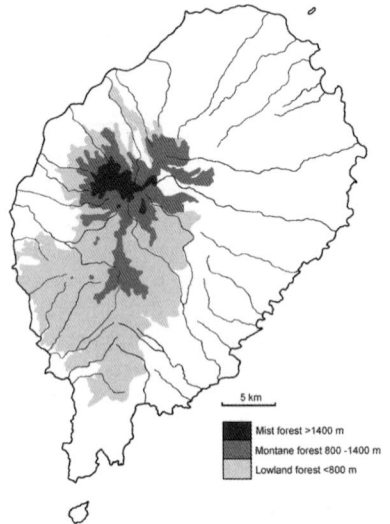

5 km

■ Mist forest >1400 m
■ Montane forest 800 -1400 m
□ Lowland forest <800 m

Figure 9 Approximate distribution of surviving primary forest on São Tomé, showing the extent of the forest zones described by Exell (1944)

(1) Lower Rainforest, extending from sea-level to 800 m, and including the endemic trees *Rinorea chevalieri*, *Chytranthus mannii*, *Anisophyllea cabole*, *Polyscias quintasii* and *Drypetes glabra*;

(2) Montane Forest (Plates 7 & 9), extending from 800 m to 1400 m, characterised by a different species composition due to lower minimum temperatures, higher rainfall and humidity, and considerable mist and cloud cover which greatly reduces light levels. The endemic trees include *Trichilia grandifolia*, *Pauridiantha insularis*, *Pavetta monticola*, *Thecacoris manniana*, *Erythrococca molleri*, *Discoclaoxylon*

Table 1   Areas of different forest types in São Tomé and Príncipe. No comparable information is available for Annobón.

	São Tomé (ha)	(%)[2]	Príncipe (ha)	ST+P (ha)	% total land area (ST+P)
Primary Forest[1] (obó)					
Lower Rainforest (0–800 m)	17,710[2]	(73%)			
Montane Forest (800–1400 m)	5787[2]	(24%)			
Mist-forest (1400–2024 m)	846[2]	(3%)			
Total	24,344[3]	4554[3]	28,898[3]	28,418[4]	28.5[4]
Secondary Forest (capoeira)	25,027[3]	3750[3]	28,777[3]	30,111[4]	30.2[4]
Shade Forest	32,289[4]	32.4[4]			

[1] Altitudinal zonation from Exell (1944).
[2] Estimated from 1:25,000 maps (Ministério do Ultramar 1958); (%) refers to percentage of total primary forest.
[3] Estimated by Jones et al (1991) from BDPA (1985).
[4] From Interforest (1990).

occidentale and *Tabernaemontana stenosiphon*. The trees are tall, forming a high, dense canopy and are covered in lianas and epiphytic mosses, ferns and orchids, the last family including the endemic species *Polystachya parviflora*, *P. ridleyi* and *Angraecum doratophyllum*. Ferns are especially abundant and diverse at these altitudes, more so than in any other part of Africa apart from Annobón (Exell 1944). Most of this vegetation zone still appears to be intact, mainly because it is at too high an altitude for successful plantation, though Snow (1950) reported encroaching cultivation near Lagoa Amélia, which continues today (Plate 10);

(3)   Mist-forest (Plate 8), from 1400 m to the top of the island at 2024 m, where temperatures are low and light is much reduced by constant mist. Endemic trees typical of these altitudes include *Podocarpus mannii*, *Balthasaria mannii*, *Psychotria guerkeana* and *P. nubicola*. At the highest altitudes the trees are small and the canopy open. Epiphytes are even more abundant and ferns remain an important element of the flora right to the summit. There is no montane grassland. Although a few alien species have been recorded in this region, perhaps owing to occupation by fugitive slaves (Chevalier 1939) and *Cinchona* spp were introduced in some parts, this habitat has been very little altered and it is nowadays seldom visited.

Estimates of the areas of different forest types are given in Table 1, though the area given for remaining primary forest in the 'Lower Rainforest' region of São Tomé is not reliable, as much of it is in fact old secondary forest on long-since abandoned plantation (Peet & Atkinson 1994).

Annobón
Exell visited Annobón briefly but relied on Mildbraed's collections made in 1911 and Wrigley's made in 1959 (Mildbraed 1922, Exell 1963). Although the island is very small, it is high enough to cast a rain-shadow to the north, where there are areas of a savanna-like formation and dry bush, including the endemic shrub *Turraea glomeruliflora*. This zone is backed on the north wall of the Quioveo crater by a dry lowland forest composed mainly of *Olea* (= *Steganthus*) *welwitschii* and *Lannea welwitschii* and containing the endemic shrubs *Maytenus annobonensis* and *Pouchetia confertiflora*. On the southern walls of the crater above 500 m and covering the south of the island to sea level grows taller mist-forest (Fig. 10), similar to the transitional montane vegetation of the summits of Príncipe. However, there is a much greater development of epiphytes, resembling the mist-forests of much higher elevations in

São Tomé. The tree *Cassipourea annobonensis* and shrub *Thecacoris annobonae* are endemic to the mist-forest of Annobón.

Although the vegetation of Annobón has to some extent been modified by man, Harrison (1990) reported that the changes are not nearly so extensive as on São Tomé and Príncipe. The north has been most affected, with plantations of bananas, jackfruit, oil-palm and mango, the latter forming quite extensive forests. There is little sign of the former plantations of cocoa and coffee, which are now abandoned and being colonised by kapok *Ceiba pentandra*, oil-palm, mango and tamarind. Most flat land within the crater is cultivated, in many places right up to the rim, though elsewhere on the crater wall the forest appears undisturbed.

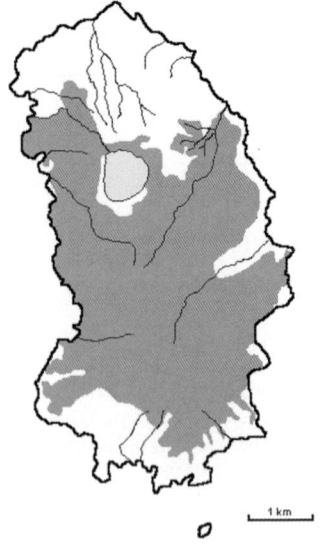

Figure 10 Approximate distribution of forest on Annobón

Endemism in the Gulf of Guinea

Floral Endemism
The floras of the islands of the Gulf of Guinea are notable for their high degree of endemism, though the numbers and frequencies of endemics recognised from each island have changed in recent years because of taxonomic revisions, as well as the discovery both of species new to the islands and of island species on the mainland (Figueiredo 1994a,b). The following figures are from Figueiredo (1994c), modifying Exell (1973).

Among the angiosperms, Príncipe has 26 single-island endemic species (8% of the total (314 spp)), São Tomé has one genus endemic to the island (*Heteradelphia*, Acanthaceae) and 81 single-island endemic species (13% of the indigenous flora (601 spp)), while Annobón has 14 (7% of the total (208 spp). Of the 176 angiosperm species endemic to the Gulf of Guinea islands overall (including 42 on Bioko), only 16 occur on more than one island (and only two of these are shared with Bioko). This emphasises the high degree of isolation under which the floras of Príncipe, São Tomé and Annobón evolved and suggests that each island received its flora separately from the mainland. The 40 single-island endemics on Bioko, by contrast, form less than 4% of its rich indigenous flora (*c* 1100 spp), as might be expected from its origin as a continental-shelf island.

Endemism among terrestrial vertebrates other than birds
Because they are oceanic islands, Príncipe, São Tomé and Annobón support few indigenous species of terrestrial vertebrates other than birds, but in all groups the levels of endemism are high.

Amphibians
Three species of amphibians occur on Príncipe and all are endemic to the Gulf of Guinea islands. Two of these also occur on São Tomé, while the other, Africa's largest tree-frog *Leptopelis palmatus*, is confined to Príncipe. The ranid previously thought to

be another single-island endemic, *Phrynobatrachus feae*, most probably refers to the non-endemic *P. dispar* (Loumont 1992). Records of a caecilian on Príncipe are now thought to be in error and that none occurs there (R. Nussbaum and local informants *in litt*, Loumont 1992, Nussbaum & Pfrender 1998).

Five species of amphibians are recognised currently on São Tomé (Loumont 1992, Nussbaum & Pfrender 1998). Three are single-island endemics: a tree-frog *Hyperolius thomensis* (formerly classified in an endemic genus *Nesionixalus*; see Drewes & Wilkinson 2004), the ranid frog *Ptychadena newtoni* and the caecilian *Schistometopum thomense*. Three species of caecilian had previously been recognised: *Schistometopum thomense, S. brevirostre* and *S. ephele* but the last two are now known to be junior synonyms of *S. thomense* (Nussbaum & Pfrender 1998). Two other species of frog, the tree-frog *Hyperolius* (formerly *Nesionixalus*) *molleri* and the ranid *Phrynobatrachus dispar*, are endemic to the two islands of São Tomé and Príncipe.

No amphibians have been reported from Annobón.

The presence of amphibians on the oceanic islands of Príncipe and São Tomé is anomalous since these animals are intolerant of salt water. The presence of a caecilian on São Tomé is particularly puzzling since it is the only known case of dispersal of this group across a marine barrier. How and when they dispersed is problematic. The fragmentation of Gondwanaland and the isolation of land-bridge islands by sea-level rise (which includes Bioko) account for all other cases of caecilian distribution (R. Nussbaum *in litt*).

Reptiles

Príncipe has eight reptiles. Five are endemic species shared with São Tomé and two, the legless skink *Feylinia polylepis* and the burrowing-snake *Typhlops elegans*, are single-island endemic species.

Fourteen species of reptiles occur on São Tomé, comprising one terrapin, eight snakes and five lizards (Haft 1993, Nill 1993, Atkinson *et al* 1994). Six of these are endemic: the snake *Philothamnus thomensis* is a single-island endemic, the gecko *Hemidactylus greeffi*, the skink *Leptosiaphos africana* and two burrowing-snakes *Rhinotyphlops feae* and *R. newtoni* are shared with Príncipe, while the gecko *Lygodactylus thomensis* is shared with Príncipe and Annobón.

Annobón supports seven species of reptiles. These include two single-island endemics *Hemidactylus newtoni* and *Leptosiaphos annobonensis*, in addition to the Gulf of Guinea endemic *Lygodactylus thomensis* that is shared with São Tomé and Príncipe (Schätti & Loumont 1992).

Mammals

When the islands were first discovered, no mammals apart from bats were reported on Príncipe, São Tomé or Annobón (Exell 1956), whereas Bioko, as a continental-shelf island, supported a large indigenous terrestrial mammal fauna (Eisentraut 1973, Fa 1991). On the three oceanic islands, all the larger mammals have been introduced since the islands' discovery, although it is not known when. São Tomé and Príncipe support populations of the monkey *Cercopithecus mona*, feral cats and pigs, the rats *Rattus rattus* and *R. norvegicus* and the mouse *Mus musculus* (Bocage 1903, 1904, Frade 1958). Annobón has *Mus musculus* (Bocage 1893b) and *Rattus norvegicus* (Basilio 1957). Two carnivores are found only on São Tomé: the African Civet *Civettictis civetta* and the large Iberian race of the weasel *Mustela nivalis numidica*. Both were introduced (Bocage 1903, Frade 1958), with the civet already having been recorded by Francisco d'Almeyda in about 1505 (Sousa 1888).

Most remarkably for such islands, however, Príncipe has an endemic subspecies of the mainland shrew *Crocidura poensis*, while São Tomé supports its own endemic species of shrew *C. thomensis* (Heim de Balsac & Hutterer 1982, Dutton & Haft 1996).

Shrews must feed every 2–3 hours and it is difficult to see how they could have survived such a sea crossing.

It is not so surprising, because of their better powers of dispersal, that bats are better represented than other terrestrial mammals on the three oceanic Gulf of Guinea islands. An analysis of the origins and species composition of the bat fauna of the Gulf of Guinea was given by Juste & Ibañez (1994). Their list is followed here.

Four species of bat have been recorded from Príncipe. *Eidolon helvum* and *Hipposideros ruber* are mainland forms, whereas a newly described subspecies of *Rousettus aegyptiacus* (*R. a. princeps*) and an undescribed new species of *Pipistrellus* are both endemic.

São Tomé supports nine species of bat. Two of these are single-island endemic species: the fruitbat *Myonycteris brachycephala* and the free-tailed bat *Chaerephon tomensis*. The fruitbat *Rousettus aegyptiacus*, the leaf-nosed bat *Hipposideros commersoni* and the long-winged bat *Miniopterus minor* are each represented by endemic subspecies on São Tomé (*R. a. thomensis*, *H. c. thomensis* and *M. n. newtoni*). The other four species, *Eidolon helvum*, *Hipposideros ruber*, *Taphozous mauritianus* and *Chaerephon pumila* are all mainland forms.

Two bats are known from Annobón, the mainland form of *T. mauritianus* and a newly described endemic subspecies, *annobonensis*, of *E. helvum* (Juste *et al* 2000).

It is interesting to compare these figures for endemism among the bats of the oceanic islands (*c* 50% on each island) with those from Bioko. Bioko supports at least 26 species of bat from seven families (much the same as Mt Cameroon), yet none appears to be endemic even at the subspecies level (Juste & Ibañez 1994).

Avian Endemism

There are difficulties in discussing the levels of avian endemism on the Gulf of Guinea islands because the various taxonomists who have made a contribution in this field have treated different taxa as distinct at the generic, species or subspecies level. We have adopted the nomenclature of *The Birds of Africa* in almost all cases but have annotated Tables 2 and 3 to show the alternative treatments proposed most recently by Sibley & Monroe (1990) and Dowsett & Dowsett-Lemaire (1993). A complete checklist of birds recorded on all four Gulf of Guinea islands is given in Appendix 1.

Out of 33 breeding landbirds, Príncipe has one monospecific endemic genus *Horizorhinus* (of uncertain affinities) and five other single-island endemic species, in addition to the five shared with São Tomé and Annobón (Table 2).

Out of a total of more than 50 breeding landbirds (the status of a handful is uncertain), São Tomé supports three monospecific endemic genera: *Amaurocichla*, *Dreptes* and *Neospiza*. There are 15 single-island endemic species and a further five endemic species that are shared with Príncipe, one of which is also shared with Annobón (Table 2).

Of the 11 species of resident landbird recorded on Annobón, *Zosterops griseovirescens* and *Terpsiphone (rufiventer) smithii* are single-island endemic species. A third, *Columba malherbii*, is shared with São Tomé and Príncipe (Table 2).

In addition, 13 mainland birds are represented by endemic subspecies on São Tomé and/or Príncipe and Annobón (Table 3).

Origins of the Avifauna

The first comprehensive discussion of the zoogeography and evolutionary origins of the Gulf of Guinea avifauna was that of Amadon (1953), based largely on J.G. Correia's collections from the late 1920s. Amadon's account is long outdated, and although subsequent authors have commented on species distributions and putative

Table 2   The endemic bird species of Príncipe (P), São Tomé (ST) and Annobón (A).

Dwarf Ibis	*Bostrychia (olivacea) bocagei*[1]	–	ST	–
São Tomé Green Pigeon	*Treron sanctithomae*[2]	–	ST	–
São Tomé Bronze–naped Pigeon	*Columba malherbii*[3]	P	ST	A
Maroon Pigeon	*C. thomensis*[4]	–	ST	–
São Tomé Scops Owl	*Otus hartlaubi*	–	ST	–
São Tomé Spinetail	*Zoonavena thomensis*	P	ST	–
Gulf of Guinea Thrush	*Turdus olivaceofuscus*			
	ssp *olivaceofuscus*	–	ST	–
	ssp *xanthorhynchus*	P	–	–
São Tomé Prinia	*Prinia molleri*	–	ST	–
Bocage's Longbill (São Tomé Short-tail)	*Amaurocichla bocagii*	–	ST	–
São Tomé Paradise Flycatcher	*Terpsiphone atrochalybeia*	–	ST	–
Annobón Paradise Flycatcher	*T. smithii*[5]	–	–	A
Dohrn's Thrush-babbler	*Horizorhinus dohrni*	P	–	–
Príncipe (Hartlaub's) Sunbird	*Anabathmis hartlaubii*	P	–	–
São Tomé (Newton's) Sunbird	*A. newtonii*	–	ST	–
Giant Sunbird	*Dreptes thomensis*	–	ST	–
Príncipe White-eye	*Zosterops ficedulinus*			
	ssp *ficedulinus*	P	–	–
	ssp *feae*	–	ST	–
Annobón White-eye	*Zosterops griseovirescens*	–	–	A
São Tomé Speirops	*Speirops lugubris*[6]	–	ST	–
Príncipe Speirops	*S. leucophaeus*	P	–	–
São Tomé (Newton's) Fiscal	*Lanius newtoni*	–	ST	–
São Tomé Oriole	*Oriolus crassirostris*	–	ST	–
Príncipe Drongo	*Dicrurus modestus*[7]	P	–	–
Príncipe Glossy Starling	*Lamprotornis ornatus*	P	–	–
Príncipe Golden Weaver	*Ploceus princeps*	P	–	–
Giant Weaver	*P. grandis*	–	ST	–
São Tomé Weaver	*P. sanctithomae*	–	ST	–
Príncipe Seedeater	*Serinus rufobrunneus*[8]			
	ssp *rufobrunneus*	P	–	–
	ssp *thomensis*	–	ST	–
	ssp *fradei*	Ilhéu Caroço	–	–
São Tomé Grosbeak	*Neospiza concolor*	–	ST	–

[1] Treated as an insular subspecies of the Olive Ibis *Bostrychia olivacea bocagei* by Sibley & Monroe (1990), Dowsett & Dowsett-Lemaire (1993) and Brown *et al.* (1982) but as an endemic species *Bostrychia bocagei* by Naurois (1973b) and Collar & Stuart (1985). We maintain its specific status (see Systematic List).

[2] Treated as conspecific with *Treron calva* by Dowsett & Dowsett-Lemaire (1993).

[3] Dowsett & Dowsett-Lemaire (1993) treated *C. malherbii*, Western Bronze-naped Pigeon *C. iriditorques* and Eastern Bronze-naped Pigeon *C. delegorguei* as conspecific.

[4] Treated as conspecific with *Columba arquatrix* by Dowsett & Dowsett-Lemaire (1993).

[5] Treated by Dowsett & Dowsett-Lemaire (1993) and Urban *et al.* (1997) as an insular subspecies of the Red-bellied Paradise Flycatcher *T. rufiventer* but we prefer to retain its specific status (see Systematic List).

[6] Considered by Dowsett & Dowsett-Lemaire (1993) to be conspecific with the Black-capped Speirops from Mt Cameroon, which was separated by Wolters (1983) as *S. melanocephalus*. Its specific separation from *S. melanocephalus* has been accepted by Fry *et al.* (2000).

[7] Considered conspecific with *Dicrurus adsimilis* by Sibley & Monroe (1990) and Dowsett & Dowsett-Lemaire (1993).

[8] Formerly in a monospecific endemic genus *Poliospiza*. Placed in *Serinus* by Sibley & Monroe (1990) and Dowsett & Dowsett-Lemaire (1993), who recognised *Poliospiza* as a subgenus.

Table 3 African mainland species with endemic subspecies on the Gulf of Guinea islands. Mainland forms of these species that also occur on one or more of the islands are given in parentheses.

	Bioko	Príncipe	São Tomé	Annobón
Olive Ibis *Bostrychia olivacea*	–	*rothschildi*	–	–
Harlequin Quail *Coturnix delegorguei*	(*delegorguei*)	–	*histrionica*	–
African Green Pigeon *Treron calva*	(*calva*)	*virescens*	–	–
Lemon Dove *Columba larvata*[1]	(*inornata*)	*principalis*	*simplex*	*hypoleuca*
Emerald Cuckoo *Chrysococcyx cupreus*[2]	(*cupreus*)	*insularum*	*insularum*	?*insularum*[3]
Barn Owl *Tyto alba*[4]	(*affinis*)	–	*thomensis*	–
Scops Owl *Otus scops*[5]	–	–	–	*feae*
Little Swift *Apus affinis*	?*bannermani*[6]	*bannermani*	*bannermani*	–
Blue-breasted Kingfisher *Halcyon malimbica*	–	*dryas*	–	–
White-bellied Kingfisher *Corythornis leucogaster*[7]	(*leucogaster*)	*nais*	–	–
Malachite Kingfisher *C. cristata*[8]	–	–	*thomensis*	–
Chestnut-winged Starling *Onychognathus fulgidus*	(*hartlaubii*)	–	*fulgidus*	–
Vitelline Masked Weaver *Ploceus velatus*	–	–	*peixotoi*	–

Following Amadon (1953) we do not accept as valid the endemic races of Laughing Dove *Streptopelia senegalensis thome*, Grey Parrot *Psittacus erithacus princeps*, Yellow-fronted Canary *Serinus mozambicus santhome*, nor the Common Waxbill *Estrilda astrild sousae* (see species accounts).

[1] *C. l. poensis* of Bioko and *C. l. hypoleuca* of Annobón were subsumed within mainland *C. l. inornata* by Fry *et al.* (1985) and Urban *et al.* (1986) but we follow Goodwin (1967) in maintaining *C. l. hypoleuca* on Annobón (see species accounts).

[2] The Emerald Cuckoo of Bioko, once considered intermediate (though unnamed) between *C. c. insularum* and mainland *C. c. cupreus*, is now considered indistinguishable from the latter (Fry *et al.* 1988).

[3] As far as we can tell, the subspecific identity of Emerald Cuckoos on Annobón has never been determined.

[4] The formerly accepted Bioko subspecies of Barn Owl, *T. a. poensis*, is now considered indistinguishable from mainland *T. a. affinis* (Fry *et al.* 1988), which should be named *T.a. poensis* by precedence (Bruce & Dowsett 2004).

[5] Dowsett & Dowsett-Lemaire (1993) regarded African *Otus scops senegalensis* as specifically distinct from Eurasian Scops Owl *O. scops* and therefore called the Annobón Scops Owl *O. senegalensis feae*.

[6] Bioko specimens appear intermediate between those from São Tomé and those from the neighbouring mainland (S. Keith *in litt*).

[7] Treated as an endemic species *Alcedo nais* by Sibley & Monroe (1990).

[8] Treated as an endemic species *Alcedo thomensis* by Sibley & Monroe (1990).

origins (eg Naurois 1983a), a more modern analysis based, for example, on molecular phylogenies is urgently required. Here we summarise what is currently known about the birds' likely ancestors and the route by which they may have arrived on the islands, whether directly from the mainland or from island to island along the archipelago.

The islands are separated from one another, and from the mainland, by only moderate distances. While Bioko is very close (32 km) to the African mainland, the three outer islands lie at increasing distances from the coast (Príncipe 220 km, São Tomé 280 km, Annobón 340 km) but are spaced at closer, similar distances (180–200 km) along the archipelago. Indeed, although neighbouring islands may not always be directly visible from the altitudes at which birds might fly, their attendant cloud cover is often visible, even from sea-level. Vagrant individuals lost over Gulf of Guinea waters might reasonably be expected to make an island landfall, whether their origin was the neighbouring mainland or another island in the archipelago.

Although we have not included the avifauna of Bioko in this Checklist, it is a possible source of colonists for the three outer islands, particularly the nearest,

Príncipe. Before its own isolation some 11,000 years ago as a continental-shelf island, Bioko would have formed a species-rich mainland peninsula with an avifauna very similar to that of nearby Mt Cameroon and its surrounding lowlands. Its geographical position would have made it a suitable springboard from which colonists might have occupied the islands serially in stepping-stone fashion. Amadon (1953) suggested that Bioko might have provided an appreciable proportion of at least the Príncipe avifauna, though he acknowledged that, because Príncipe is about equidistant from Bioko and the much more extensive mainland to the east, either source seems equally likely to have provided colonists. In a few cases, he suggested that the absence of a species from Bioko indicated that colonisation of the outer islands took place directly from the mainland. Bioko's present-day avifauna, however, is considerably impoverished by comparison with the neighbouring mainland. Although almost all the expected bird families are still represented, it may have lost a number of species that were unable to sustain viable populations after isolation. As a result, Bioko might nowadays be expected to supply fewer potential colonists for the outer islands, and tracing the origins of island birds by examining current distributions on Bioko and the mainland must be done with caution.

For the same reason, the present-day avifauna of Príncipe may also be impoverished compared with the prehistoric past. Bathymetric evidence suggests that, of all the three oceanic islands, Príncipe has varied the most in surface area with fluctuations in sea-level during the last 20,000 years. Until c 12,000 years ago, its land area would have been almost ten times larger than at present (1179 km^2 v 128 km^2 - larger even than present-day São Tomé), and extending to include the Ilhas Tinhosas 25 km to the south (Juste & Ibañez 1994). Príncipe may therefore have accommodated many more species than its present greatly reduced size can support, removing evidence of further lineages once shared with Bioko or the other islands.

Species with an island-hopping origin

Amadon thought that the present-day distributions of three species indicated a colonisation of Príncipe via Bioko, rather than from the mainland directly. Two of these are not supported by present evidence: the Emerald Cuckoo *Chrysococcyx cupreus*, which could have come from either source, and the Olive Sunbird *Cyanomitra olivacea* (discussed later). Only the third of Amadon's examples, the Little Swift, plus the speiropses, which he only briefly considered, do indeed seem to be linked to the mainland via Bioko:

(1)   Little Swift *Apus affinis*: subspecies *bannermani* is endemic to Príncipe and São Tomé, and is a clear case where Príncipe shares a lineage with Bioko rather than with the mainland. Bioko specimens appear intermediate between those from São Tomé and those from the neighbouring mainland (A.T. pers obs, S. Keith *in litt*, see also Amadon 1953).

(2)   Speiropses *Speirops* spp: the three endemic island species of the zosteropid genus *Speirops* probably represent an early invasion of the Gulf of Guinea islands. The ancestors of *S. brunneus* (Bioko), *S. leucophaeus* (Príncipe) and *S. lugubris* (São Tomé) may have colonised the archipelago by island-hopping from the mainland, where the fourth member of the genus, *S. melanocephalus*, is endemic to Mt Cameroon. Alternatively, it is possible that the genus originated on São Tomé and secondarily invaded the mainland (Moreau 1957, Fry *et al* 2000).

Species originating directly from the mainland

Only one species present on Príncipe, and with close relatives on Bioko, seems to owe its origin to the mainland rather than Bioko: the Olive Sunbird *Cyanomitra olivacea*. Amadon (1953) believed that the Olive Sunbird was represented by the same subspecies *C. o. obscura* on Bioko and Príncipe but it has now been shown that the

Príncipe race is *C. o. cephaelis* from the mainland (Tye & Macaulay 1993), which indicates a direct, and perhaps relatively recent, colonisation.

Other birds present on Príncipe, São Tomé or Annobón are also likely to have had a direct mainland origin because they are absent from Bioko, though it should be remembered that a present-day absence from Bioko does not mean that it was always so.

(1)   Olive ibises *Bostrychia* spp: absent from Bioko, the Olive Ibis *Bostrychia olivacea* has differentiated into endemic taxa both on Príncipe and São Tomé. It is represented by an endemic race *rothschildi* on Príncipe and a closely related endemic species, the Dwarf Ibis *Bostrychia bocagei* confined to São Tomé. The Dwarf Ibis presumably derived from a separate, earlier colonisation but there is no evidence to indicate whether its affinities are closer to mainland or Príncipe birds.

(2)   São Tomé Bronze-naped Pigeon *Columba malherbii*: this species is endemic to Príncipe, São Tomé and Annobón, forming a superspecies with the Western Bronze-naped Pigeon *C. iriditorques* of the adjacent mainland (Urban *et al* 1986). It is not clear, however, whether the colonisation proceeded as an initial invasion to one of the islands that subsequently spread to the other two, or as three independent colonisations. Whatever the case, it is curious that neither it nor the Western Bronze-naped Pigeon is present on Bioko.

(3)   The Blue-breasted Kingfisher *Halcyon malimbica* is represented on Príncipe by the endemic subspecies *H. m. dryas* but is absent from Bioko, São Tomé and Annobón.

(4)   Príncipe Golden Weaver *Ploceus princeps*: this appears to have no close relatives, though it shows similarities with both the Cape Weaver *P. capensis* of South Africa and Holub's Golden Weaver *P. xanthops* of the nearby Gulf of Guinea mainland.

Direct colonisation from the African mainland would seem even more likely when the species is also absent from Príncipe as well as Bioko, though again it must be remembered that Príncipe may have lost a significant proportion of its ice-age avifauna as sea-levels rose and its surface area diminished. Some of the following species may therefore have had an island-hopping origin, of which the intermediate forms have been lost.

(1)   Harlequin Quail *Coturnix delegorguei*: subspecies *histrionica* is confined to São Tomé but the species is an intra-African migrant. It is known from only a single vagrant record on Bioko and is absent from Príncipe, suggesting a direct mainland origin.

(2)   Scops owls *Otus* spp: there is no scops owl on Bioko although there may be an undescribed one on Príncipe. It seems likely that the São Tomé Scops Owl *O. hartlaubi* and the Annobón Scops Owl *O. scops feae* both derive from colonists that came directly from the mainland.

(3)   Malachite Kingfisher *Corythornis cristata*: subspecies *C. c. thomensis* is endemic to São Tomé. The species is unknown on Bioko and its niche on Príncipe is occupied by an endemic race of the White-bellied Kingfisher *C. leucogaster nais*.

(4)   The São Tomé Prinia *Prinia molleri* was placed by Hall & Moreau (1970) within the White-chinned Prinia *P. leucopogon* superspecies but the most recent treatment offers no suggestions as to its closest ally (Urban *et al* 1997). The absence of any other prinias from the Gulf of Guinea islands suggests that the São Tomé Prinia had a direct mainland origin.

(5)   São Tomé Paradise Flycatcher *Terpsiphone atrochalybeia*: confined to São Tomé, this forms a superspecies with the African Paradise Flycatcher *T. viridis* of the

mainland and the endemic island species of the Seychelles, Mascarenes, Comoro Is and Madagascar (Urban *et al* 1997). There is no paradise flycatcher at all on Príncipe and the taxon is represented on Bioko, not by *T. viridis*, but an endemic race of the Red-bellied Paradise Flycatcher *T. rufiventer tricolor*.

(6)   Annobón Paradise Flycatcher *T. smithii*: clearly derived from the Red-bellied Flycatcher *T. rufiventer* rather than *T. viridis* and accorded only subspecific status by Urban *et al* (1997). The remoteness of Annobón and the absence of close relatives on Príncipe and São Tomé suggest that the Annobón Paradise Flycatcher colonised this tiny island directly from the mainland rather than by island-hopping from Bioko. Given its less complete differentiation from its mainland ancestor and uncertain specific status compared to the São Tomé Paradise Flycatcher, it would seem to be of a more recent origin.

(7)   São Tomé (Newton's) Fiscal *Lanius newtoni*: a member of the *L. collaris* superspecies of the African mainland. Since no other indigenous shrikes are known from either Bioko or Príncipe, this species presumably originated directly from the mainland. It could be argued, however, that this forest-dwelling fiscal lives at such low densities that the intermediate island populations may have been unable to survive isolation.

(8)   The São Tomé Oriole *Oriolus crassirostris* (confined to São Tomé) is thought to be closest to the Western Black-headed Oriole *O. brachyrhynchus* (Naurois 1984d), which does not occur on either Bioko or Príncipe. The only oriole on Bioko is the Black-winged Oriole *O. nigripennis*.

(9)   Vitelline Masked Weaver *Ploceus velatus*: subspecies *P. v. peixotoi* confined to São Tomé.

Species of uncertain origin

The routes by which endemic taxa colonised the oceanic islands remain uncertain where their closest relatives (other subspecies or members of the same superspecies) occur both on Bioko and the surrounding Gulf of Guinea mainland. The origins of others are even more difficult to determine where there are no close relatives nearby, and especially so where their taxonomic affinities are also in doubt.

(1)   The green pigeons *Treron* spp: the nominate form of the African Green Pigeon *Treron c. calva* occurs both on Bioko and the African continent surrounding the Gulf of Guinea, so that the endemic Príncipe race *T. c. virescens* could have originated from either source. Its relative, the São Tomé Green Pigeon *T. sanctithomae*, could therefore have originated via Bioko-Príncipe stock or have come directly from the mainland.

(2)   Lemon Dove *Columba* (*Aplopelia*) *larvata*: we have followed Goodwin (1967) in considering the Lemon Dove to have differentiated subspecifically on each of the four islands (eg Goodwin 1967), a pattern that could reflect an island-hopping origin from Bioko. More recently, however, *C. l. poensis* of Bioko and *C. l. hypoleuca* of Annobón have been subsumed within mainland *C. l. inornata* (Fry *et al* 1985, Urban *et al* 1986), leaving the Bioko and Annobón races more closely related to each other and mainland birds than to those on the two middle islands of the archipelago, and suggesting a more complex pattern of colonisation.

(3)   Maroon Pigeon *Columba thomensis* (São Tomé): this forms a superspecies with the Cameroon Olive Pigeon *C. sjostedti* (Bioko and Cameroon highlands) and the Olive (Rameron) Pigeon *C. arquatrix* of northwest Angola, southern and eastern Africa (Urban *et al* 1986), though these three are sometimes treated as conspecific within *C. arquatrix* (Dowsett & Dowsett-Lemaire 1993). Naurois (1988a) considered the Maroon Pigeon to show a greater resemblance to

mainland *C. arquatrix* than to *C. sjostedti* of Bioko which, together with its absence from Príncipe, argues against an island-hopping origin. However, the nearest present-day population of *C. arquatrix* (*sensu stricto*) is in northwestern Angola, well distant from the Gulf of Guinea islands and unlikely nowadays to provide a source of potential colonists. This superspecies, all members of which live in high altitude forests, may represent relicts whose lowland ancestor has disappeared. Alternatively, this could be a case where the population on Príncipe has died out since the last ice age after an island-hopping origin.

(4) Emerald Cuckoo *Chrysococcyx cupreus*: Bioko birds were previously thought to be intermediate between mainland *C. c. cupreus* and the race *C. c. insularum*, which is endemic to São Tomé, Príncipe and (possibly) Annobón (Moreau & Chapin 1951). Bioko specimens are now considered to be indistinguishable from mainland birds (Eisentraut 1973, Fry *et al* 1988), so the origin of *insularum* remains uncertain, while the population on Annobón has never been properly described.

(5) Barn Owl *Tyto alba*: Barn Owls are absent from Príncipe, and the supposedly darker Bioko birds, once separated as *T. a. poensis*, are now included within mainland *T.a. affinis* (Fry *et al* 1988, see also Bruce & Dowsett 2004), so it remains uncertain whether the dark endemic São Tomé race *T. a. thomensis* came via Bioko or directly from the mainland.

(6) São Tomé Spinetail *Zoonavena thomensis* (Príncipe and São Tomé): the other two African species of *Zoonavena* do not occur on the mainland but on the Comoro Islands and Madagascar and presumably have no recent connection with the Gulf of Guinea species, perhaps indicating that '*Zoonavena*' is polyphyletic. The genus is most closely related to the *Raphidura* spinetails, of which the sole African species, Sabine's Spinetail *R. sabini*, occurs on Bioko and the lowland forests of the Gulf of Guinea coast.

(7) White-bellied Kingfisher *Corythornis leucogaster*: The nominate form occurs both on Bioko and the mainland, so that the endemic race *C. l. nais* on Príncipe (sometimes considered specifically distinct as *Alcedo nais* (Sibley & Monroe 1990)) could equally well be derived from either source.

(8) Gulf of Guinea Thrush *Turdus olivaceofuscus* (Príncipe and São Tomé): considered to be older than, and not closely related to, any African mainland thrush, including the African Thrush *T. pelios* of Bioko and the West African mainland; possibly derived from the now extinct mainland ancestor of both *T. pelios* and the Olive Thrush *T. olivaceus* (Keith & Urban 1992, Urban *et al* 1997).

(9) Bocage's Longbill (São Tomé Short-tail) *Amaurocichla bocagii*: placed *incertae sedis* among the Timaliidae by Dowsett & Forbes-Watson (1993) but more firmly allied with the warblers by Urban *et al* (1997), its affinities remain a matter of conjecture (see species accounts).

(10) Dohrn's Thrush-babbler *Horizorhinus dohrni*: placed *incertae sedis* among the Timaliidae by Dowsett & Forbes-Watson (1993), its affinities remain a matter of conjecture (see species accounts).

(11) Príncipe (Hartlaub's) Sunbird *Anabathmis hartlaubii* and São Tomé (Newton's) Sunbird *A. newtonii*: regarded by Amadon (1953) as belonging to the *Anabathmis* subgenus of *Cyanomitra* sunbirds. The genus *Anabathmis* is now recognised to contain only three species: the third is Reichenbach's Sunbird *A. reichenbachii* of the West African mainland; no members of the genus occur on Bioko (Fry *et al* 2000). These two island sunbirds might therefore have derived either directly from the mainland, or possibly via a now-extinct ancestral population on Bioko.

(12) Giant Sunbird *Dreptes thomensis* (São Tomé): formerly grouped with Príncipe (Hartlaub's) and São Tomé (Newton's) Sunbirds in the *Anabathmis* subgenus of *Cyanomitra* and later subsumed in *Nectarinia* (see above), the original monospecific genus *Dreptes* has now been reinstated, reflecting this bird's greater differentiation from its apparent *Anabathmis* ancestor (Fry *et al* 2000). Presumably the Giant Sunbird represents an earlier colonisation by the *reichenbachii-hartlaubii-newtonii* lineage but it is not now possible to trace the route by which this happened (Fry *et al* 2000).

(13) White-eyes *Zosterops* spp: there is a species of *Zosterops* on each of the Gulf of Guinea islands, that of Bioko being a race *poensis* of the mainland Yellow White-eye *Z. senegalensis*, while São Tomé and Príncipe each have a race of the endemic *Z. ficedulinus* and Annobón has the endemic *Z. griseovirescens*. Their pattern of colonisation was probably similar to that of the islands' earlier occupation by *Speirops*, by island-hopping at least from Príncipe, though Príncipe itself could have been initially colonised from either Bioko or the mainland.

(14) Príncipe Drongo *Dicrurus modestus*: the Príncipe Drongo is very close to the Fork-tailed Drongo *D. adsimilis* and is often considered conspecific with it. *D. adsimilis coracinus* occurs throughout the Gulf of Guinea mainland and on Bioko, so either source could have provided potential colonists for Príncipe.

(15) Príncipe Glossy Starling *Lamprotornis ornatus*: a mainland origin for the Príncipe Glossy Starling would seem to be indicated by analogy with the subspecific identity of its sister species, the Splendid Glossy Starling *L. splendidus*, which is at least sporadically present on Príncipe. Splendid Glossy Starlings occur both on Bioko and the Gulf of Guinea mainland, so birds could potentially arrive from either direction. However, the invaders are mainland *L. s. splendidus* rather than the endemic Bioko race *L. s. lessoni*, which may be evidence that the Príncipe Glossy Starling derived from similar but much earlier invasions of the ancestral stock directly from the mainland. Alternatively, however, it could be argued that the ancestral glossy starling arrived from Bioko before that island's recent isolation gave rise to its own endemic subspecies.

(16) Chestnut-winged Starling *Onychognathus fulgidus*: absent from Príncipe; the nominate subspecies is endemic to São Tomé, while *O. f. hartlaubii* occupies both Bioko and the West African mainland.

(17) Giant Weaver *Ploceus grandis*: this is derived from the Village Weaver *P. cucullatus*, which occurs on both Bioko and São Tomé itself but is absent from Príncipe. There is no evidence to suggest whether the ancestor of the Giant Weaver came via Bioko rather than directly from the mainland. The presence of Village Weavers on São Tomé probably represents a recent secondary invasion which could also have occurred either from Bioko or directly from the mainland, or could even be a human introduction.

(18) São Tomé Weaver *Ploceus sanctithomae*: the São Tomé Weaver was included, with reservations, by Hall and Moreau (1970) in a superspecies with the Brown-capped Weaver *P. insignis*, which occurs on Bioko, and Preuss's Golden-backed Weaver *P. preussi* from the mainland.

(19) Príncipe Seedeater *Serinus rufobrunneus*: formerly treated as belonging to an endemic monotypic genus *Poliospiza* but included by Fry & Keith (2004) within subgenus *Crithagra*, not their expanded subgenus *Poliospiza*. Its affinities with any mainland species are unclear, though both Amadon (1953) and Naurois (1975b) point out its similarities to the Thick-billed Seedeater *Serinus burtoni* of East Africa. Whether the initial colonisation happened via Bioko or directly

from the mainland, island-hopping subsequently occurred between Príncipe and São Tomé.

(20) São Tomé Grosbeak *Neospiza concolor*: while having many similarities to the ploceine genus *Amblyospiza*, this monospecific endemic genus is now generally agreed to be a fringillid closest to *Poliospiza* (= *Serinus*) and probably represents an earlier invasion of the islands by the same ancestral stock that later gave rise to the seedeater (Amadon 1953, Naurois 1975b, 1988c). Like the colonisation by the *Anabathmis-Dreptes* sunbirds, it is not now possible to trace the route by which this occurred.

Summary

Amadon's suggestion that Bioko was a probable source for the initial colonists of the three outer islands in the Gulf of Guinea is only partially supported by present-day evidence. Of the 19 endemic taxa found on Príncipe, only two (11%) show unequivocal evidence of an origin via Bioko (Table 4a), whereas five (26%) seem likely to have arrived directly from the mainland. In the case of the remainder, it is not possible to decide: seven (37%) could have come either from Bioko, the mainland directly, or secondarily via São Tomé; the remaining five (26%) represent more ancient colonisations whose nearest affinities are obscure.

An origin via Bioko seems to have occurred even less often for the endemic taxa on São Tomé. Of the 28 taxa listed in Table 4b only two (7%) clearly have an island-hopping origin, whereas 10 (36%) seem to have come directly from the mainland. Nine others (32%) have identifiable relatives on Bioko and may have arrived by island-hopping via Príncipe, or directly from the mainland, or from the mainland via Príncipe. The remaining seven species (25%) are of older, uncertain origin. Amadon thought it probable that Príncipe served as a springboard for the double invasions of São Tomé by the *Dreptes-Anabathmis* sunbirds and by the *Neospiza-Poliospiza* finches, but neither lineage is clearly derived initially from Bioko rather than the mainland directly.

Because of Annobón's geographical location closer to São Tomé than to the mainland, it is not unexpected that a majority of its restricted avifauna should have arrived by island-hopping along the archipelago rather than directly from the mainland. Four of its six endemic taxa almost certainly island-hopped from São Tomé but both the Annobón Scops Owl and Annobón Paradise Flycatcher seem to have had a direct mainland origin (Table 4c). The scops owl is only subspecifically separated from the mainland form and the specific status of the paradise flycatcher is debatable, whereas their São Tomé relatives are well differentiated and clearly not ancestral.

Given the archipelago's long evolutionary history, its present-day avifauna offers only a fragmentary glimpse of the possible scenarios that gave rise to it. It remains to be seen whether the relationships suggested by traditional morphological and behavioural comparisons are supported by the molecular taxonomy that is now required.

Bird Habitats

Forest and other terrestrial habitats

Before their relatively recent colonisation and modification by man, the islands were covered mainly by tropical rainforest, with only very limited areas of other habitats such as dry forest or mangrove. Presumably, therefore, most of the indigenous avifauna will have evolved as forest birds. Only in the last few hundred years has the range of habitats available to the birds increased, although the greater part of the land surface of all three islands remains under some form of forest cover. Only a relatively small part of this, however, remains suitable for the endemic species of most conservation concern.

Table 4 Suggested origins and affinities of the endemic birds of (a) Príncipe, (b) São Tomé, and (c) Annobón in relation to their occurrence on Bioko and the mainland. In each case, taxa shared with other islands are indicated in parentheses by P = Príncipe; ST = São Tomé; A = Annobón.

(a) Príncipe	Island-hopping via Bioko	Direct from mainland	Uncertain origin
Endemic subspecies	intermediate form on Bioko Little Swift (ST)	other ssp on Bioko Olive Sunbird species absent from Bioko Olive Ibis Blue-breasted Kingfisher São Tomé Bronze-naped Pigeon (ST, A)	other ssp on Bioko African Green Pigeon Lemon Dove (ST, A) Emerald Cuckoo (ST, A) White-bellied Kingfisher Príncipe Drongo
Endemic species	sister-species on Bioko & mainland Príncipe Speirops	sister-species on Bioko & mainland – no close relative on Bioko Príncipe Golden Weaver	sister-species on Bioko & mainland Príncipe White-eye (ST) Príncipe Glossy Starling no close relative on Bioko São Tomé Spinetail (ST) Gulf of Guinea Thrush (ST) Príncipe (Hartlaub's) Sunbird Príncipe Seedeater (ST)
Endemic genus			Dohrn's Thrush-babbler

(b) São Tomé	Island-hopping	Direct from mainland	Uncertain origin
Endemic subspecies	intermediate on Bioko same ssp on Príncipe Little Swift (P)	species absent from Bioko São Tomé Bronze-naped Pigeon (P, A) species absent from Bioko & Príncipe Harlequin Quail Malachite Kingfisher Vitelline Masked Weaver	other ssp on Bioko Lemon Dove (P, A) Emerald Cuckoo (P, A) Barn Owl Chestnut-winged Starling
Endemic species	sister-species on Príncipe, Bioko & mainland São Tomé Speirops	sister-species on Príncipe, Bioko & mainland – no close relative on Bioko Dwarf Ibis São Tomé Scops Owl São Tomé Paradise Flycatcher no close relative on Bioko or Príncipe São Tomé Prinia São Tomé Oriole São Tomé (Newton's) Fiscal	sister-species on Bioko & mainland São Tomé Green Pigeon Maroon Pigeon Príncipe White-eye (P) Giant Weaver São Tomé Weaver no close relative on Bioko São Tomé Spinetail (P) Gulf of Guinea Thrush (P) São Tomé (Newton's) Sunbird Príncipe Seedeater (P)
Endemic genera			Bocage's Longbill (São Tomé Short-tail) Giant Sunbird São Tomé Grosbeak

(c) Annobón	Island-hopping	Direct from mainland
Endemic subspecies	other ssp on Bioko Lemon Dove (P, ST) Emerald Cuckoo (P, ST) species absent from Bioko São Tomé Bronze-naped Pigeon (P, ST)	species absent from Bioko, Príncipe & São Tomé Annobón Scops Owl
Endemic species	sister-species on Bioko, Príncipe & São Tomé Annobón White-eye	sister-species on Bioko & mainland only Annobón Paradise Flycatcher

Bird habitat choice has been best studied on São Tomé (Peet & Atkinson 1994). Primary forest and mature secondary forest both appear to support almost exclusively endemic birds, with rather more present in primary forest at low altitudes compared with higher elevations and secondary forest (Table 5a). Six species are primary forest specialists and have not so far been recorded in mature secondary forest. Four of these (Dwarf Ibis, São Tomé (Newton's) Fiscal, Bocage's Longbill (São Tomé Short-tail) and São Tomé Grosbeak) have not been recorded above 600 m, so would appear to be confined to Exell's (1944) 'Lower Rainforest' region (<800 m; see Vegetation). Given that much of the forest below 800 m is old secondary forest, the area of suitable remaining habitat for these species is much smaller than the 177 km² suggested in Table 1. This lower rainforest zone also supports the two other primary forest specialists, the Giant Sunbird and Maroon Pigeon, and contains the greater part of the populations of São Tomé Scops Owl, São Tomé Green Pigeon, São Tomé Oriole, Giant Sunbird and the São Tomé race of the Príncipe White-eye. Indeed, all the endemic species on São Tomé occur in lowland primary forest and no other habitat supports populations of every one of them.

No species is confined to the higher altitude montane and mist-forest zones, although the Maroon Pigeon has its stronghold there. Most of the other endemic species are also common at high altitude but the green pigeon, bronze-naped pigeon, scops owl, paradise flycatcher, oriole, and Giant Weaver all have their altitudinal limit at about 1600 m and are not found in the Pico de São Tomé area.

With the exception of the six primary forest specialists mentioned above, all the endemic birds also occur in secondary forest. Mature secondary forest provides an important habitat for the São Tomé Scops Owl, Gulf of Guinea Thrush and São Tomé Oriole, but seems to be especially important for the Giant Weaver, which is a secondary forest specialist. The small remaining areas of dry forest in the north of São Tomé support a few of the most widespread endemic species (eg São Tomé (Newton's) Sunbird, São Tomé Speirops, São Tomé Oriole and São Tomé Weaver) but none has a substantial part of its population there.

Table 5 Numbers of endemic and non-endemic breeding birds found in different habitats on Príncipe and São Tomé (modified from Peet & Atkinson 1994).

(a) Príncipe

	Primary forest[1]	Secondary forest	Plantation	Rivers and wetlands
Endemic species	11	7	7	0
Endemic subspecies	7	7	6	1
Non-endemic	2	2	10	5
Totals	20	16	23	6

[1] Primary forest on Príncipe resembles the lowland forest of São Tomé and there is no clearly defined montane forest region.

(b) São Tomé

	Montane and mist-forest	Lower rainforest	Secondary forest	Plantation	Dry savanna	Rivers and wetlands
Endemic species	16	20	16	12	12	1
Endemic subspecies	2	3	4	7	7	1
Non-endemic	0	0	0	6	18	6
Totals	18	23	20	25	37	8

Plantations, whose vegetation structure often falls between that of natural forest and man-altered habitats, contains populations of both endemic and non-endemic birds. Several endemic species have adapted well to cocoa plantations, where the retention of shade trees produces a forest structure of upper and lower storeys. The extensive areas of cocoa plantations may support significant populations of São Tomé Paradise Flycatcher, São Tomé (Newton's) Sunbird, São Tomé Speirops, São Tomé Prinia, Gulf of Guinea Thrush and Giant Weaver.

A few of the more widespread endemic birds may be seen in town gardens, such as the São Tomé Sunbird and the São Tomé Prinia. The most conspicuous species in grassland, farmland, gardens and along roadsides, however, are non-endemic, such as the Laughing Dove *Streptopelia senegalensis* and the various weavers and widowbirds. It is possible that many, if not all, of these species have arrived naturally or been introduced over the past five centuries and have become established only as a result of habitat modification by man.

Habitat choice among birds on Príncipe (Table 5b) and Annobón is much less well known. The scarce Príncipe races of the Gulf of Guinea Thrush and Príncipe White-eye both appear to be confined to primary forest. The Príncipe Seedeater also occurs almost exclusively in primary forest, though it may use long-abandoned coconut plantations on the south coast of the island, and is numerous on the Ilhéu Caroço (Boné de Jóquei). The other endemic species are encountered readily in cocoa plantations, farmland and along roadsides, though the Príncipe Drongo is scarce.

The two endemic species on Annobón, the Annobón Paradise Flycatcher and Annobón White-eye, are both common in cultivated areas, secondary forest and the moist forest at higher altitude (Fry 1961, Harrison 1990); the paradise flycatcher is reported as far commoner than its São Tomé counterpart (Harrison 1990). Among the other landbirds, the Lemon Dove, São Tomé Bronze-naped Pigeon and the Annobón Scops Owl are confined to the forest, where the last species is apparently common.

Wetland and coastal habitats

Marshy habitat is limited to tiny areas along some rivers, most of which are scoured and stony (Plates 21 & 22). On river estuaries there are a few small areas of *Pandanus* (Plate 26) and mangroves, the largest of which is Malanza in the south of São Tomé (Plate 19). Most mangroves are badly damaged by cutting for fuelwood but some remain suitable as nesting sites for Reed Cormorants *Phalacrocorax africanus* and herons; a big colony of Cattle Egrets *Bubulcus ibis* uses the remaining mangrove at Agua Izé. The only endemic birds associated with these wetlands are the two subspecies of *Corythornis* kingfishers but these are also widespread along streams in forest. Tidal mudflats and sandy shorelines are likewise very restricted (Plates 20 & 25). No resident species are associated with them and they support Palearctic migrant shorebirds in only small numbers.

Marine habitats

The seabird communities of the Gulf of Guinea islands are much less well known than the forest avifauna, mainly because most ornithological visitors to the islands had less interest in seabirds than in the unknown birds of the forest interior, and so they did not visit the various small offshore islets. Even now, not all have been visited during the breeding seasons of the various species, so the breeding population sizes of most seabirds remain unknown.

The only breeding seabird known from the main island of São Tomé is the White-tailed Tropicbird *Phaethon lepturus*, which is abundant breeding not only in cavities in coastal cliffs but also in tree cavities several kilometres inland and at up to 500 m altitude (Naurois 1973a). Breeding of a possibly distinct race of Madeiran Storm-Petrel *Oceanodroma castro* has not so far been established, despite occasional records

of this species and unidentified storm-petrels inland (Amadon 1953, Atkinson *et al* 1994, A. Gascoigne pers comm).

The biggest concentration of breeding seabirds offshore from São Tomé is on the Sete Pedras, a group of rocky islets *c* 5 km off the SE coast (Plates 33 & 34), which support a few tens of pairs of White-tailed Tropicbirds and a colony of about a hundred Brown Boobies *Sula leucogaster* (Naurois 1973a, Atkinson *et al* 1994). A small number of Bridled Terns *Sterna anaethetus* may also breed there (Naurois 1973a). Several other islets off the São Tomé coast also support small numbers of tropicbirds, with a few tens of pairs present on the two biggest, Ilhéu das Cabras (2 km from the NE coast) and Ilhéu de Santana (1 km from the E coast).

The main island of Príncipe appears to support no breeding seabird species, not even the White-tailed Tropicbird that is so conspicuous on São Tomé. Tropicbirds nest on two of the islets close inshore, however, with significant colonies on Ilhéu Caroço (*c* 3 km off the SE corner; Plates 27 & 28) and Ilhéus dos Mosteiros (*c* 1 km from NE corner); Ilhéu Caroço also has a colony of 100+ Brown Boobies (Naurois 1973a, Monteiro *et al* 1997).

Without doubt the most important seabird breeding colony in the Gulf of Guinea is the Ilhas Tinhosas, 25 km SSW of Príncipe (Plates 31 & 32). These comprise two small islands, Tinhosa Grande (*c* 20 ha) and Tinhosa Pequena (*c* 3 ha), which, in the 1960s, supported several hundred breeding pairs of Brown Boobies, 0.6–1.2 million pairs of Sooty Terns *Sterna fuscata* and 20,000–50,000 pairs each of Common and Black Noddies *Anous stolidus* and *A. minutus* (Naurois 1973a). The most recent survey, in July 1997 (outside the peak breeding period for most species and therefore yielding minimum figures), has estimated 1500–3000 Brown Boobies, 111,000 Sooty Terns, 10,000–20,000 Common Noddies and 4000–8000 Black Noddies (Monteiro *et al* 1997). The Ilhas Tinhosas must therefore rank among the most important seabird sites in the eastern tropical Atlantic.

Seasonality of Breeding

Breeding patterns

There are only fragmentary data on breeding for many Gulf of Guinea species. In some cases, breeding has been inferred from the birds' behaviour or, in the case of collected specimens, from gonad condition. In other cases there are confirmed records of nest-building, eggs or young but rarely in sufficient quantity to give a true indication of the length of the breeding season.

In the following analyses we have grouped birds by feeding guild and simply totalled the number of species recorded laying in each month. In some of these there were enough data to analyse São Tomé and Príncipe separately but there were insufficient data for Annobón, which has been excluded. The full data are given under the individual species accounts in the Systematic List.

Laying by most terrestrial insectivores begins with the onset of the rains between August and October, at the end of the main dry period, and finishes just after the shorter dry period in January; the pattern is similar on the two islands (Fig. 11). It is not known what seasonal changes may occur in the biomass of invertebrate prey and we can only speculate why insectivores breed during one wet period rather than the other. The shorter dry period in January is probably of relatively little significance in terms of reduced insect food abundance compared to the longer dry period in July and August, though it may signal the end of breeding and the start of the post-nuptial moult. It is probably advantageous for the moult to be completed before the start of the major dry season in July.

There is less information for frugivores, which are here taken to include the doves, parrots and starlings; data for the two islands are combined in Figure 12. Breeding occurs between October and February, a pattern rather similar to that of insectivores. The timing of breeding among frugivores may be explained if the seasonal pattern of fruit availability follows that on the nearby mainland. In Gabon, fruit abundance is unimodal, increasing from October onwards, declining in April–May and becoming scarce in June–August (White 1994, Tutin & White 1998). If this is true also on São Tomé, there may therefore be better breeding opportunities for frugivorous birds immediately after the main dry season ends.

The breeding seasons of seedeaters on São Tomé seem much more prolonged, though fewer lay in February–March (Fig. 13). Some species, such as Laughing Dove, Giant Weaver and Bronze Mannikin *Lonchura cucullata* may breed throughout the year. On Príncipe the breeding season of seedeaters appears more confined to the middle of the year but limited data for only three species are involved. The more prolonged breeding seasons of granivorous birds probably reflects a more constant availability of seeds throughout the year, though many of these species also require green grass for nest-building and insect food when breeding. Perhaps the only part of the year that

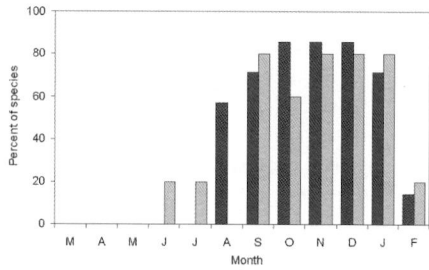

Figure 11 Laying dates of insectivorous birds on Príncipe (grey bars) and São Tomé (black bars). Species included are (Príncipe, *n* = 5): Blue-breasted Kingfisher, White-bellied Kingfisher, Dohrn's Thrush-babbler, Príncipe (Hartlaub's) Sunbird, Príncipe Speirops; (São Tomé, *n* = 7): Emerald Cuckoo, Gulf of Guinea Thrush, São Tomé Prinia, São Tomé Paradise Flycatcher, São Tomé (Newton's) Sunbird, Giant Sunbird, Príncipe White-eye.

would be consistently unsuitable for breeding by many weavers is the dry season, especially where it is most marked in the savanna grassland of northern São Tomé, and such species might be expected to show greater seasonality there than in the southern, wetter areas.

The breeding seasons for aerial feeders (swifts) are quite different from those of other insectivores. The three species of swift on São Tomé breed between April and October, peaking during the drier period in June to September (Fig. 14). There are no data for swifts on Príncipe other than the Little Swift, which breeds rather earlier, in April and May, than it does on São Tomé, where it breeds from June to September. Whether this is a genuine difference in seasonality between the two islands, or

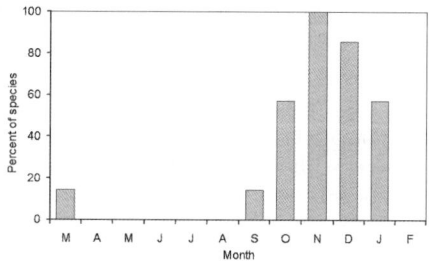

Figure 12 Laying dates of frugivorous birds on Príncipe and São Tomé combined. Species included are: African Green Pigeon, São Tomé Green Pigeon, São Tomé Bronze-naped Pigeon, Lemon Dove, Grey Parrot, Red-headed Lovebird, Príncipe Glossy Starling. Note that the year runs from March to February

simply reflects a lack of adequate data, remains unknown. Nothing is known of seasonal variations in insect abundance, so the difference in timing of breeding between aerial feeders and the other insectivores remains unexplained.

The limited data for three species of herons (Cattle Egret, Green-backed Heron *Butorides striatus* and Western Reef Heron *Egretta gularis*) suggest a long, almost year-round, breeding season from August until June, with perhaps a peak in October–January and some differences in timing between São Tomé and Príncipe (see Systematic List). Reed Cormorants have been reported to breed in association with Cattle Egrets and Reef Herons, with eggs laid in November. A Dwarf Ibis with enlarged gonads in November would also fit with a similar breeding schedule.

Breeding records for seabirds have been summarised by Naurois (1973a) and Monteiro *et al* (1997) but our knowledge remains incomplete. There appear to be two peak periods, corresponding to the

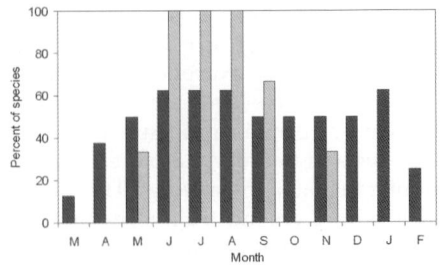

Figure 13 Laying dates of seed-eating birds on Príncipe (grey bars) and São Tomé (black bars). Species included are (Príncipe, *n* = 3): Príncipe Seedeater, Príncipe Golden Weaver, Red-headed Quelea; (São Tomé, *n* = 8): Harlequin Quail, Laughing Dove, Yellow-fronted Canary, Príncipe Seedeater, Vitelline Masked Weaver, Giant Weaver, São Tomé Weaver, Bronze Mannikin. Note that the year runs from March to February

two dry seasons of December–January and June–August (Fig. 15). However, while some species are well synchronised, the breeding seasons of others are protracted and their duration unknown, perhaps even following sub-annual cycles as is documented for some species elsewhere. The Black Noddy on Tinhosa Grande has a well-synchronised breeding season that includes the dry season, laying after April and fledging young shortly after July (mainly between August and October on Annobón; Fry 1961). Common Noddies are also well-synchronised but with peak laying probably taking place between July and September. Sooty Terns have two peaks a year, in December–January and May–June, or sometimes October–November, though whether the breeding cycle is sub-annual at *c* 9 months, as suggested by Naurois (1973a), is unknown. Bridled Terns have also been recorded laying in October and November. White-tailed Tropicbirds and Brown Boobies both have protracted seasons with laying occurring from November–December to May–June.

There is insufficient information available to speculate on the factors determining the breeding seasons of freshwater species or seabirds.

Migration in the Gulf of Guinea

Because very few ornithologists have spent much time in the Gulf of Guinea islands, migrants that are uncommon, or occur only on passage, are likely to be overlooked. Nevertheless, the scarcity of such records suggests that few Palearctic migrants use Príncipe, São Tomé or Annobón with any regularity and that virtually no Afrotropical migrants do so.

Among Palearctic migrant waders, only the Common Sandpiper *Actitis hypoleucos* is common on all four islands while Whimbrels *Numenius phaeopus* are common on the oceanic islands but sporadic on Bioko. Among Palearctic seabirds, only Sandwich *Sterna sandvicensis* and Common/Arctic Terns *S. hirundo/paradisaea* are seen frequently

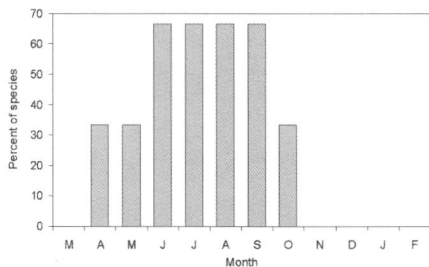

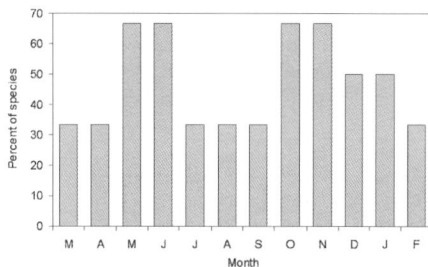

Figure 14 Laying dates of aerial feeders in Príncipe and São Tomé combined (São Tomé Spinetail, Little Swift, Palm Swift). Note that the year runs from March to February

Figure 15 Laying dates of seabirds on Príncipe and São Tomé. Species included are White-tailed Tropicbird, Brown Booby, Sooty Tern, Bridled Tern, Black Noddy, Common Noddy. Note that the year runs from March to February

inshore (Table 6). Most Palearctic landbirds overwintering in West Africa are western European species that favour the savannas north of the rainforest zone; only Yellow Wagtail *Motacilla flava* and Tree Pipit *Anthus trivialis* are regular on Bioko, within sight of the mainland. Even fewer reach the outer islands, where the only Western Palearctic birds that seem to be regular in small numbers are Barn Swallow *Hirundo rustica* and perhaps Great Reed Warbler *Acrocephalus arundinaceus*, which are trans-equatorial migrants. Other species that overwinter in southern Africa, however, such as European Roller *Coracias garrulus*, House Martin *Delichon urbicum*, Spotted Flycatcher *Muscicapa striata*, Garden Warbler *Sylvia borin*, Lesser Grey Shrike *Lanius minor*, Red-backed Shrike *L. collurio* and European Golden Oriole *O. oriolus*, are known only as vagrants in the Gulf of Guinea islands. Apart from Garden Warbler and Golden Oriole, none of these has been recorded from Bioko (Peréz del Val 1996). Records are so few that it seems probable that Palearctic migrants ordinarily do not venture out across the Gulf of Guinea but remain over the mainland.

The occurrence of Afrotropical migrants is just as rare and sporadic (Table 7). Again, none is common on Bioko; a few occur there regularly in small numbers or as vagrants but most are unknown on the outer three islands. Others that have been recorded as vagrants on the outer islands are either not known even from Bioko (eg Great White Egret *Ardea alba*, Purple Heron *A. purpurea*) or are only vagrants there too (Grey Heron *A. cinerea*). The occasional occurrence of rails and gallinules, which are such highly dispersive species, is not surprising, but even birds that are not known as migrants on the African mainland also turn up on the islands from time to time (Black Heron *Egretta ardesiaca*, Yellow-billed Stork *Mycteria ibis*).

Only the Cattle Egret is a regular migrant between October and May on Annobón and here it is not certain whether the birds originate from the mainland or from the breeding population on São Tomé. It is not known whether Red-headed Queleas *Quelea erythrops*, which have bred on São Tomé and formerly did so on Príncipe, are sedentary local residents or breeding migrants, or indeed whether they are irregular immigrants that manage to sustain a local population for short periods. In many such cases it would be impossible to distinguish island residents from continental immigrants.

While more systematic observations by ornithologists resident in the islands throughout the year will undoubtedly yield more records of migrants, it already seems clear that the Gulf of Guinea islands are not on any major migration route, and

Table 6 Occurrence of Palearctic migrants on the Gulf of Guinea islands. M = regular migrant in moderate numbers; m = probably a regular migrant in small numbers; V = vagrant, recorded <5 times; ? = status uncertain or unconfirmed report. Data for Bioko are from Peréz del Val (1996).

	Bioko	Príncipe	São Tomé	Annobón
White Stork *Ciconia ciconia*			V	
Common Kestrel *Falco tinnunculus*		?		
Western Red-footed Falcon *F. vespertinus*			V	
Peregrine Falcon *F. peregrinus*	V		V	
Black-winged Pratincole *Glareola nordmanni*		V		V
Little Ringed Plover *Charadrius dubius*	V		?	
Ringed Plover *C. hiaticula*	m		?	
'Lesser' Golden Plover *Pluvialis sp.*			V	
Eurasian Golden Plover *P. apricaria*			?	
Grey Plover *P. squatarola*	V		V	
Sanderling *Calidris alba*				m
Pectoral Sandpiper *C. melanotos*		V		
Curlew Sandpiper *C. ferruginea*	V	V	V	
Common Snipe *Gallinago gallinago*	V			
Bar-tailed Godwit *Limosa lapponica*		V	V	V
Whimbrel *Numenius phaeopus*	V	M	M	M
Eurasian Curlew *N. arquata*		V	V	
Common Redshank *Tringa totanus*	V			
Marsh Sandpiper *T. stagnatilis*	m			
Greenshank *T. nebularia*	V	m	m	
Green Sandpiper *T. ochropus*	V		?	
Wood Sandpiper *T. glareola*	V		m	
Common Sandpiper *Actitis hypoleucos*	M	M	M	
Ruddy Turnstone *Arenaria interpres*		V	V	
Pomarine Skua *Stercorarius pomarinus*			V	
Arctic Skua *S. parasiticus*		?	?	?
Long-tailed Skua *S. longicaudus*		?	?	?
Lesser Black-backed Gull *Larus fuscus*		?	?	
Sabine's Gull *L. sabini*		?	?	?
Royal Tern *Sterna maxima*	V	?	?	
Sandwich Tern *S. sandvicensis*			m	
Common Tern *S. hirundo*	V	?	?	
Arctic Tern *S. paradisaea*	V	?	?	
Whiskered Tern *Chlidonias hybrida*	V			
Black Tern *C. niger*	M	?	?	
White-winged Black Tern *C. leucopterus*				V
Common Swift *Apus apus*			?	
Alpine Swift *Tachymarptis melba*				m
European Roller *Coracias garrulus*		V	V	
Barn Swallow *Hirundo rustica*	m	V	m	
House Martin *Delichon urbicum*		V		
Yellow Wagtail *Motacilla flava*	M			
Tree Pipit *Anthus trivialis*	M		?	
Whinchat *Saxicola rubetra*		V		
Great Reed Warbler *Acrocephalus arundinaceus*			V	
Willow Warbler *Phylloscopus trochilus*	V	?	?	
Wood Warbler *P. sibilatrix*	V			
Garden Warbler *Sylvia borin*	V		V	V
Spotted Flycatcher *Muscicapa striata*	V	V	V	V
Lesser Grey Shrike *Lanius minor*		V		V
Red-backed Shrike *L. collurio*			V	V
European Golden Oriole *Oriolus oriolus*	V	V		

Table 7   Afrotropical species recorded as regular migrants or vagrants on the Gulf of Guinea islands. R = resident (given only for species that occur as migrants on another of the islands); M = well-established migrant; m = migrant in small numbers only; (br) = migrant known to breed; V = vagrant (5 records or fewer); E = extinct; ? = uncertain status. Status on Bioko from Peréz del Val (1996) and Pérez del Val *et al* (1997) (correcting Dowsett & Dowsett-Lemaire 1993).

	Bioko	Príncipe	São Tomé	Annobón
White-breasted Cormorant *Phalacrocorax carbo*		?	?	
Reed Cormorant *P. africanus*	V		R	
Darter *Anhinga rufa*	V			
Black-crowned Night Heron *Nycticorax nycticorax*			?[1]	
Squacco Heron *Ardeola ralloides*	V[1]	?[1]		
Cattle Egret *Bubulcus ibis*	M[1]	R?	R?	M?
Green-backed Heron *Butorides striatus*	R	R	R	V
Black Heron *Egretta ardesiaca*			V	
Little Egret *E. garzetta*	M[1]	?	?	
Great White Egret *Ardea alba*			?	
Purple Heron *A. purpurea*			V[1]	
Grey Heron *A. cinerea*	V[1]	V[1]	V[1]	
Yellow-billed Stork *Mycteria ibis*			V	
Lesser Flamingo *Phoeniconaias minor*		V		
Knob-billed Duck *Sarkidiornis melanotos*			V	
Black (Yellow-billed) Kite *Milvus migrans* (ssp *parasitus*)	M	R	R	V
African Fish Eagle *Haliaeetus vocifer*	V			
African Crake *Crex egregia*	V	V	V	
African Water Rail *Rallus caerulescens*			V	
Allen's Gallinule *Porphyrio alleni*	M	V	V(br?)	V
Lesser Moorhen *Gallinula angulata*	V	V	R?	
White-fronted Sand Plover *Charadrius marginatus*	V		V	
Jacobin Cuckoo *Oxylophus jacobinus*		V	V	
Great Spotted Cuckoo *Clamator glandarius*		?		V[1]
Pennant-winged Nightjar *Macrodipteryx vexillaria*	m?			
Black Swift *Apus barbatus*	V?	?	?	
Senegal Kingfisher *Halcyon senegalensis*	M?			
Giant Kingfisher *Megaceryle maxima*	V			
Pied Kingfisher *Ceryle rudis*	R	R?	V	
Banded Martin *Riparia cincta*		V		
Grey-rumped Swallow *Pseudhirundo griseopyga*		?		
Lesser Striped Swallow *Hirundo abyssinica*	m(br)			
Red-headed Quelea *Quelea erythrops*	V	E/V?	R?	
Red-billed Quelea *Q. quelea*			?	
Long-tailed Paradise-Whydah *Vidua paradisaea*			?	
Yellow-fronted Canary *Serinus mozambicus*			R	?

[1] It is not known if these were Palearctic or Afrotropical migrants.
[2] Possibly escaped cage birds.

that the majority of migrants, whether Palearctic or Afrotropical, remain tied to the African mainland.

Bird Species Turnover

The brief analysis above shows that the Gulf of Guinea islands sporadically receive vagrant individuals of mainland species that could potentially establish breeding populations. By definition, those that are classified here as vagrants have not established themselves, yet other species appear to have colonised the islands as breeding birds in the very recent past. Some idea of what changes may have occurred can be gained from the early collections, not only from what the collections contain but also from what they do not. This assumes that the early collectors made diligent attempts to collect everything. From their journals and from the prevailing ethos of the times, it would seem that they did, so that gaps in their collections are very likely to represent genuine absences.

Species turnover on Príncipe

There appear to have been some extinctions on Príncipe as well as new colonisations (Table 8). At some time between 1866 and 1901, both the Red-headed Lovebird *Agapornis pullaria* and the Red-headed Quelea disappeared. Both were found by Dohrn and Keulemans in 1865 but neither was collected by Fea in 1901, nor have they been recorded since. However, it is not certain that the lovebird was ever established as a breeding species (see Systematic List).

Two species may have disappeared and then recolonised within the last 100–150 years. The Black (Yellow-billed) Kite *Milvus migrans parasitus* was reported to visit the south coast of Príncipe in 1865 (Keulemans 1866), but it was then apparently absent until after Snow's 1949 visit; it was recorded again in 1954 by Frade. Likewise, the Common Waxbill *Estrilda astrild* was rare in 1865 and thereafter seems to have been absent (eg Frade & Vieira dos Santos 1977) until re-established sometime recently (see Systematic List). The latter may represent two human-mediated introductions.

Apart from the kite and waxbill, four other species have become established on Príncipe in the 20th century. The Palm Swift *Cypsiurus parvus* did not colonise the island until some time after Correia's departure in 1929 but was present by 1949 (Snow 1950). Snow (1950) thought that the Laughing Dove was introduced around 1935. Neither Cattle Egret nor Common Moorhen *Gallinula chloropus* were recorded until 1954 and 1956, respectively (Frade & Vieira dos Santos 1977).

If we ignore the two recolonists (kite and waxbill), the turnover rate over the period in question was 9% (for method of calculation see footnote to Table 8). If the two extinctions and re-colonisations are included in the calculation, the species turnover rate on Príncipe has been 16%. The natural species turnover rate would have been much lower if some of these populations derived from escaped cage birds.

Species turnover on São Tomé

With the recent rediscoveries of the Dwarf Ibis, São Tomé (Newton's) Fiscal, Bocage's Longbill (São Tomé Short-tail) and the São Tomé Grosbeak, all of São Tomé's birds recorded by 19th-century collectors still survive. In addition, however, ten species that do not appear in earlier collections have become established as breeding birds during the past century or so (Table 8).

The earliest 'new' arrival was the Fire-crowned Bishop *Euplectes hordeaceus*, which was first recorded in 1893 and was initially uncommon. This was followed in 1900 by the first record of Laughing Dove from a collection made available to Bocage by A. Negreiros (Bocage 1904); it had not been recorded by Newton prior to that. Next was the White-winged Widowbird *Euplectes albonotatus*, first recorded by Alexander in 1909. The Southern Cordon-bleu *Uraeginthus angolensis* and Village Weaver *Ploceus cucullatus* were collected by Correia in 1928 but had not been recorded by Boyd Alexander or anyone before him. A later arrival on São Tomé was the Palm Swift. Frade did not record it there in 1954, although it had already arrived on Príncipe by 1949 (Snow 1950), and it was first seen on São Tomé in 1959 (Fry 1961). Three other species are more puzzling. The Lesser Moorhen *Gallinula angulata* was not collected by Correia and was therefore thought by Naurois (1983a) to have colonised after 1929, but 19th-century specimens suggest its earlier, perhaps sporadic, occurrence. Another is the Grey Parrot *Psittacus erithacus*, first reliably recorded on São Tomé in

Table 8   Colonisations and extinctions on the oceanic islands of the Gulf of Guinea over the past 100–150 years. Status on other islands indicated by: R = long-established resident population; V = vagrant/migrant in small numbers; – = not recorded.

	Príncipe	São Tomé	Annobón
Cattle Egret	first recorded 1954	R	V
Bubulcus ibis			
Western Reef Heron	R	R	first recorded 1909
Egretta gularis			
Black (Yellow-billed) Kite	vagrant in 19th C;	R	V
Milvus migrans	established by 1954		
Red-necked Spurfowl		first recorded 1987	
Francolinus afer			
Common Moorhen	first recorded 1956	R	extinct after 1910
Gallinula chloropus			
Lesser Moorhen	V	?vagrant in 19th C;	–
G. angulata		established after 1929?	
Laughing Dove	?introduced *c* 1935	first recorded 1900	–
Streptopelia senegalensis			
Grey Parrot	R	established by 1972	–
Psittacus erithacus			
Red-headed Lovebird	extinct after 1865	R	–
Agapornis pullaria			
Palm Swift	first recorded 1949	first recorded 1959	–
Cypsiurus parvus			
Vitelline Masked Weaver	–	first recorded 1928	–
Ploceus velatus			
Village Weaver	–	first recorded 1928	–
P. cucullatus			
Red-headed Quelea	extinct after 1865	R/(migrant?)	–
Quelea erythrops			
Fire-crowned Bishop	–	first recorded 1893	–
Euplectes hordeaceus		?introduced	
White-winged Widowbird	–	first recorded 1909	–
Euplectes albonotatus			
Common Waxbill	extinct after 1865;	R	–
Estrilda astrild	re-established after 1949		
Southern Cordon-bleu	∫	first recorded 1928	–
Uraeginthus angolensis		?introduced	
Bronze Mannikin	R	R	first recorded 2000
Lonchura cucullata			?introduced
Turnover rate[1]	16%	11%	15%

[1] Apparent species turnover rate during the period n is calculated as $T_n = (E_n + I_n)/(S_y + S_{y+n})$, where E_n and I_n are the extinction and immigration rates, and S_y and S_{y+n} are the numbers of species present at the beginning and end of the period in question (Russell *et al* 1995).

1972 after earlier unsubstantiated reports of transient birds (Naurois 1983b; see Systematic List), and which now occurs in small scattered groups (Nadler 1993). The other is the Vitelline Masked Weaver *Ploceus velatus*. The first record seems to be that of Correia as late as 1928–29, with only one doubtful specimen before that, and Bannerman (1915a) pointed out that no-one had recorded it in the then recent past. This example is puzzling because the species is represented by an apparently well-defined endemic race *peixotoi*. However, subspecific differentiation does not rule out the possibility that the bird is a recent arrival (and possibly an introduction, see

Frade & Naurois 1964), and may simply reflect a founder effect deriving from a small initial gene pool.

If all ten species are genuinely new colonists, this represents a 22% increase in the number of breeding landbirds originally recorded for São Tomé and an apparent species turnover rate of c 11% over the past 100–150 years. As on Príncipe, however, many of these new species are likely to have been kept as caged birds and may therefore have established from escapees. If this were so, the natural species turnover rate would be considerably reduced.

Species turnover on Annobón

Only 11 breeding species other than seabirds have been recorded from Annobón. Of these, three were introduced: the Domestic Chicken *Gallus gallus* was present by about 1500 (Monod *et al* 1951), the Helmeted Guineafowl *Numida meleagris* was established before 1846 (Allen & Thomson 1848), and the Bronze Mannikin *Lonchura cucullata*, presumably deriving from escaped cage birds, has established very recently (Pérez del Val 2001). In addition, there seem to have been two natural changes in the avifauna in the past hundred years or so. The Common Moorhen was last recorded in 1910 (Lowe 1932) but was said to have gone extinct only a few years before 1955 (Basilio 1957), while the Western Reef Heron, a normally conspicuous bird which had not been recorded by either Newton in 1892–93 or Fea in 1902, was first seen by Alexander in 1909. The Black (Yellow-billed) Kite, which has established itself on Príncipe and São Tomé within historical times, has been recorded as a vagrant on Annobón but has not so far established itself there. Thus, with one extinction and two colonisations, the species turnover rate on Annobón has been 15% over the past 100–150 years.

Summary

The majority of the colonists on all three islands are grain-eating birds of open and disturbed areas (dove, weavers, waxbills), while the Black (Yellow-billed) Kite is a generalised scavenger, the Palm Swift has colonised coconut plantations and the remainder are waterbirds; only the Grey Parrot is a bird of forested habitats. What is surprising is that some of these species apparently did not colonise earlier, despite the fact that suitable habitat, especially for granivores, must have been available to them for several hundred years after the initial opening up of the forest for cultivation. Some weavers and waxbills were already present at an early date, though recently enough to have remained undifferentiated from mainland forms. Like the recent colonists, however, it is unknown which might have arrived naturally or been deliberately or accidentally introduced; the granivorous species are among the most likely to have originated from escaped cage birds. The failure of other vagrants to establish (eg the waterbirds in Table 7) must reflect either a lack of sufficient suitable habitat to sustain a viable population, or simply too few individuals arriving simultaneously.

The Future of the Avifauna

Threats

The political isolation of the Gulf of Guinea islands within post-colonial Africa came to an end in the late 1980s, exposing heavily degraded infrastructures and near-subsistence economies that were woefully inadequate for their future development. Food production was poorly developed, the cash crop economy based on cocoa, coffee and copra had suffered from falling world prices and years of neglect, and São Tomé e Príncipe and Equatorial Guinea numbered amongst the poorest countries in the world. Although change has been rapid in the last decade, there remain the inevitable problems for sustainable development posed by expanding human populations confined to small, geographically isolated oceanic islands (IUCN 1993). This section briefly discusses some of the factors that are likely to affect the future of bird populations on the Gulf of Guinea islands, some of which have been operating for decades or centuries past, and others that will inevitably accompany economic development.

Human population growth

In 2000, the population of São Tomé & Príncipe was estimated as c 160,000 and increasing at 3.16% pa (CIA 2001). Recent figures are not available separately for Bioko and Annobón but in 1983 the population of Bioko was c 59,000 and Annobón c 2000 (Fa 1991). The population growth rate in Bioko is unknown but is likely to be at least as high as in São Tomé & Príncipe, given its apparently low level in the early 1980s on what is a large and fertile island. The relative stability and slow population growth on Annobón (the population was c 1300 at the beginning of the 20th century) has been due to continued emigration to Bioko and the mainland (Fa 1991). In the face of these rapidly increasing human populations, it is not obvious how the islands' economic future and ecological integrity can be jointly sustained.

Habitat change

During the latter part of the present century, the major habitat change in São Tomé and Príncipe was the reversion of many plantations to secondary regrowth as estates were abandoned, a process that accelerated after independence in 1975 (Plates 23 & 24). The changes have been greatest in former cocoa and coffee plantations on richer soils in wetter areas. Many emergent forest trees had been retained in plantations to provide shade for cocoa bushes and several exotic shade species had been planted, notably *Erythrina* spp (Plate 12), so that abandoned plantations now carry tall, dense secondary forest (*capoeira*). The extent of this regrowth is considerable, covering 30% of the land area of São Tomé and Príncipe combined (Table 1) and providing a significant increase in the amount of habitat suitable for many forest birds. In contrast, the abandonment of coconut plantations has merely resulted in the growth of a dense understorey of herbaceous plants and creepers that has not greatly increased the attractiveness of these areas for birds; coconut plantations remain one of the most species-poor habitats on the islands. In some southern areas of São Tomé, profusely regenerating oil-palm *Elaeis guineensis* is rapidly becoming an invasive weed and may threaten the integrity of the primary forest.

The major habitat modification likely to occur in São Tomé and Príncipe in the near future will be a loss of *capoeira* as the abandoned plantations are reclaimed, either to rehabilitate the cocoa industry if prices improve, or to promote timber and fuelwood production (Interforest 1990). There is already renewed pressure on timber resources for house construction accompanying privatisation and redistribution of land. Agricultural development and changes in land-use patterns will also undoubtedly have some consequences for bird populations. Most of the commoner endemic birds survive well in cocoa plantations managed in the traditional manner under large shade trees (Jones & Tye 1988, Atkinson *et al* 1991, Peet & Atkinson 1994). It is a matter of great concern, however, that under modern management, which does not require shade trees, this habitat could be severely impoverished if fallen shade trees are not replaced or the existing ones removed.

Much of the remaining virgin rainforest on São Tomé and Príncipe has been given some legal protection as *zonas ecológicas* and these are included within the 299km^2 Parque Natural Ôbo de São Tomé e Príncipe, approved by the National Assembly in Dec 2003. (see below, Conservation Initiatives). Nevertheless, a recent proposal to develop a Free Zone area in southwestern Príncipe gave cause for concern

(GGCG 1996c, 1997). The area of the concession would have overlapped with a large part of the *zona ecológica* that is intended to protect Príncipe's primary forest. The idea was abandoned, but not for conservation reasons.

Pesticides

It has been claimed that increased use of pesticides to improve cocoa production in the years immediately before independence caused significant declines in the populations of some insectivores, notably the São Tomé Paradise Flycatcher (Naurois 1984a) and possibly also the white-eye and oriole. To some extent these species recovered as agriculture fell into decline after independence but they and others may prove vulnerable once more to increasing pesticide use as agricultural production re-intensifies and public health programmes tackle the problems of disease control.

Hunting and trade

Small birds are commonly hunted for food by children with catapults but a cause for concern is the hunting of the endemic pigeons on São Tomé and Príncipe by those with access to firearms and ammunition (mainly workers on the large agricultural enterprises, the *empresas*). Particularly vulnerable is the Maroon Pigeon (Jones & Tye 1988, Atkinson *et al* 1994).

Some species, mostly waxbills and parrots, are captured as local pets and for the international bird trade but data on the extent of this problem are largely lacking. No endemic species appears to be traded. Trade in parrots appears to have had little effect on their populations in the past, except that it may have helped cause the extinction of the Red-headed Lovebird *Agapornis pullaria* as a breeding bird on Príncipe (but see Systematic List). Since the 1980s, however, the capture of Grey Parrots *Psittacus erithacus* on Príncipe has taken place on a commercial scale and local people have already reported a noticeable decline in the population (Juste 1996). Various estimates have been made of the numbers of adults and chicks taken but there are no data on what level of harvesting might be sustainable in the long term. Harrison & Steele's (1989) report of 3000 chicks taken annually is likely to be an overestimate, while two more recent estimates of numbers harvested each year gave 1500 chicks plus an unknown number of adults (Juste 1996), and 430–550 chicks plus 200–300+ adults (Melo 1998). Although all parrot trappers interviewed claimed that they would cease harvesting if they had alternative sources of income (Melo 1998), the financial incentive for trade in parrots may increase if the island's traditional economy based on cocoa and coffee production collapses further, since a parrot hunter can earn double the mean annual income in only two months (Juste 1996).

A significant threat to the Gulf of Guinea seabird community is the disruption to the breeding colonies on the Ilhas Tinhosas caused by fishermen collecting large numbers of Brown Booby chicks for sale as food (Christy 1995b). Traditionally, fishermen have collected eggs and young birds for domestic use but the availability of new glass-fibre canoes with outboard motors and ice compartments for storage has enabled them to remain in the vicinity of the islets for several days, so that a commercial harvest has developed. Particularly at risk is the Brown Booby colony on the plateau of Tinhosa Grande, the only area accessible to the fisherman, but the disturbance threatens all species breeding on the islets. Adult and young terns are also taken for use as bait.

Introduced predators

As on so many oceanic islands that lacked indigenous land mammals, introduced mammalian predators on the Gulf of Guinea islands have almost certainly had an adverse effect on native bird faunas that evolved in their absence (see review by Dutton 1994). Most of the introduced mammals arrived centuries ago, however, and their impact on the original avifauna, or indeed their effects today, can only be guessed at. The introductions inevitably included man's commensals, the Ship (Black) Rat *Rattus rattus*, Norway (Brown) Rat *R. norvegicus* and House Mouse *Mus musculus*. The Ship Rat would have been one of the first alien predators to arrive on the islands in the late fifteenth century and would have had an immediate impact, particularly because of its tree-climbing ability, which would have enabled it to reach almost any bird's nest (Atkinson 1985). The Brown Rat would not have reached the Gulf of Guinea until after 1700 (Atkinson 1985) but, since then, both species will have remained important nest predators. Three other naturalised introductions will also have had a significant effect as opportunistic predators of birds and their nests: Mona Monkey *Cercopithecus mona*, African Civet *Civettictis civetta* and European Weasel *Mustela nivalis*. Domestic and feral cats, dogs and pigs are also likely to have a significant impact on bird populations; there have been reports of wild dogs taking eggs and young of ibises (Collar & Stuart 1985).

Species at risk

The endemic birds can be grouped into three categories according to the extent to which they may be at risk (Table 9). The majority of endemic birds are common and under no immediate threat on any of the three islands (Table 9a). The reason for this is that many of them appear to have adapted well to plantations and other man-made habitats. This conclusion needs to be treated with some caution, however, because there are no data on breeding success in different habitats. It may well be that some species are common in secondary habitats yet breed poorly there compared with those in undisturbed forest, their populations being maintained by surplus birds immigrating from areas where they reproduce well. Such data are urgently needed. Two species in Table 9a, the Annobón Paradise Flycatcher and Annobón White-eye, are considered to be 'Vulnerable' by BirdLife International (2000), despite having apparently adapted well to man-made habitats, but this classification reflects the tiny size of the island rather than any immediate threat.

The second category comprises species that appear to do well at present in managed habitats but it is uncertain what intensity of management they can tolerate; several have been listed as 'Vulnerable' by BirdLife International (Table 9b). On São Tomé the removal of shade trees from plantations has been shown to cause the Gulf of Guinea Thrush to disappear and populations of several other species to decrease, including the São Tomé Paradise Flycatcher, São Tomé (Newton's) Sunbird, São Tomé Speirops and São Tomé Prinia (Atkinson *et al* 1994). The flycatcher was said to have decreased after 1971 as a result of massive insecticide use in the cocoa plantations, while populations in forest and abandoned cultivation remained unchanged (Naurois 1984a). Pesticides were much less widely used after 1975 and by the late 1980s paradise flycatchers were once again common in plantations with shade trees (Jones & Tye 1988, Atkinson *et al* 1991, Peet & Atkinson 1994). None of these species is currently thought to be threatened, though any felling of shade trees to satisfy the demand for timber, or intensification of pesticide use associated with rehabilitation of the plantations, could affect them adversely.

A similar decline was reported in the early 1970s for populations of the Príncipe White-eye on both islands (Naurois 1983a, Collar & Stuart 1985). The São Tomé subspecies appeared subsequently to have recovered in numbers (Jones & Tye 1988, Harrison & Steele 1989) Atkinson *et al* (1991) but recent visitors have found its range to remain restricted (P. Kaestner *in litt*). The Príncipe subspecies has remained rare (Jones & Tye 1988) and both subspecies are listed as 'Vulnerable'.

The Príncipe Seedeater had also become rare on Príncipe by the late 1980s (Jones & Tye 1988) and was not seen by Atkinson *et al* (1991), although the subspecies on São Tomé is common and ubiquitous. Naurois (1975b) thought its populations underwent marked natural fluctuations but, whatever the reasons for its rarity, the ability of the Príncipe race to recover from low population levels may be reduced by increased

Table 9. The status of the endemic birds of Príncipe, São Tomé and Annobón according to their present habitat use and likely tolerance of future changes in land use. Endemic species are in bold, endemic subspecies are in normal type. Threatened status of endemic species according to BirdLife International is given as follows: CR = Critically Endangered; VU = Vulnerable; NT = Near Threatened; LC = Least Concern; subspecies are not categorised[1] (Stattersfield *et al* 1998, BirdLife International 2000).

(a) Common species using man-altered habitats; not immediately threatened.

Príncipe	Status	São Tomé	Status
African Green Pigeon	–	Harlequin Quail	–
São Tomé Bronze-naped Pigeon	LC	**São Tomé Green Pigeon**	LC
Lemon Dove	–	**São Tomé Bronze-naped Pigeon**	LC
Emerald Cuckoo	–	Lemon Dove	–
Little Swift	–	Emerald Cuckoo	–
São Tomé Spinetail	LC	Barn Owl	–
Blue-breasted Kingfisher	–	**São Tomé Spinetail**	LC
White-bellied Kingfisher	LC[1]	Little Swift	–
Dohrn's Thrush-babbler	LC	Malachite Kingfisher	LC[1]
Príncipe (Hartlaub's) Sunbird	LC	**São Tomé Prinia**	LC
Príncipe Glossy Starling	LC	São Tomé (Newton's) Sunbird	LC
Príncipe Golden Weaver	LC	**São Tomé Speirops**	LC
		Chestnut-winged Starling	–
Annobón		Vitelline Masked Weaver	–
Emerald Cuckoo	–	**Giant Weaver**	LC
Annobón Paradise Flycatcher	VU	São Tomé Weaver	LC
Annobón White-eye	VU	**Príncipe Seedeater**	LC

[1] BirdLife International recognises the São Tomé and Príncipe races of the Malachite and White-bellied Kingfishers as full species and has therefore assigned a category of threat.

(b) Species using man-altered habitats but vulnerable to agricultural improvements.

Príncipe	Status	São Tomé	Status
Gulf of Guinea Thrush	NT	**Gulf of Guinea Thrush**	NT
Príncipe White-eye	VU	São Tomé Paradise Flycatcher	LC
Príncipe Speirops	NT	**Príncipe White-eye**	VU
Príncipe Drongo	NT		
Príncipe Seedeater	LC		

Annobón
São Tomé Bronze-naped Pigeon[2]   LC
Lemon Dove[2]   –

[2] There is some confusion and information is lacking about the relative abundance and vulnerability of the two pigeons on Annobón.

(c) Forest species intolerant of land-use change and largely restricted to undisturbed habitats.

Príncipe	Status	São Tomé	Status
Olive Ibis	–	**Dwarf Ibis**	CR
		Maroon Pigeon	VU
Annobón		**São Tomé Scops Owl**	VU
Annobón Scops Owl	–	**São Tomé (Newton's) Fiscal**	CR
		São Tomé Oriole	VU
		Bocage's Longbill (São Tomé Short-tail)	VU
		Giant Sunbird	VU
		São Tomé Grosbeak	CR

habitat disturbance. The endemic race of the seedeater on the Ilhéu Caroço is still abundant (M. Melo pers comm) and presumably still safe.

Three further birds may be at risk on Príncipe: the endemic subspecies of the Gulf of Guinea Thrush, Príncipe Speirops and Príncipe Drongo. All occur in disturbed habitats but at lower density than most of the other endemic birds, and they may be especially susceptible to further habitat change or increased pesticide use (Jones & Tye 1988, Atkinson et al 1991). Given the small size of Príncipe and the low density of birds, the speirops, thrush and drongo (currently 'Near Threatened') come close to meeting the criteria for the 'Vulnerable' category (BirdLife International 2000).

The third category in Table 9 comprises those birds of primary rainforest that use man-altered habitats very little, and then only when these have been abandoned for a long time and have largely reverted to forest. Some of these species may never have been common, or became rare following the agricultural expansions of the nineteenth century. Seven were included in the bird Red Data Book (Collar & Stuart 1985) but, in the latest listing by BirdLife International of threatened species worldwide, 11 were included as 'Critically Endangered' or 'Vulnerable' and three more as 'Near Threatened' (BirdLife International 2000). All have their strongholds in the remaining areas of undisturbed forest, where their total populations are likely to be small. Three of the four recently rediscovered species on São Tomé were listed by BirdLife as 'Critical'. Of these, the Dwarf Ibis might require a large home range so that its population would necessarily be small on a small island, while the São Tomé Grosbeak and São Tomé (Newton's) Fiscal have been seen so seldom that their populations are probably very small. None of these has been seen in any habitat other than primary forest and, in recent times, only the fiscal has been seen as high as c 600 m altitude (it was collected by Correia at 1060 m). If they do occur at higher altitudes their populations might be in the hundreds to low thousands; if they do not, their populations might be only in the tens or low hundreds (Peet & Atkinson 1994). Bocage's Longbill (São Tomé Short-tail) is considered 'Vulnerable' (BirdLife International 2000), with an estimated surviving population of 680–930 birds (Peet & Atkinson 1994). If, as seems to be the case, the ibis, longbill, and grosbeak are restricted to altitudes below 500 m, then only the forest in the Xufexufe and Ana Chaves valleys are large enough to support viable populations, and these would be precariously small (Peet & Atkinson 1994).

The Maroon Pigeon and Giant Sunbird, both listed as 'Vulnerable' (BirdLife International 2000), have their strongholds in the high altitude primary forest above 1400 m on São Tomé, where their undoubtedly small populations would be gravely threatened by development of their restricted habitats. Also considered 'Vulnerable' on São Tomé is the São Tomé Oriole, which, although still quite common in primary and older secondary forest, has not recolonised cocoa plantations where it occurred in the early 1970s before the use of pesticides (Naurois 1984d).

Conservation initiatives

The global importance of the Gulf of Guinea islands for the conservation of biodiversity was recognised by ICBP in the early 1980s, when Collar & Stuart (1988) ranked the rainforests of São Tomé and Príncipe second out of 75 African forests in their importance as bird habitats, based on the numbers of their endemic species and their level of vulnerability. BirdLife International now includes Príncipe, São Tomé and Annobón among more than 200 Endemic Bird Areas (EBAs) worldwide (Bibby et al 1992, Stattersfield et al 1998) and all three islands are included among the Important Bird Areas (IBAs) of Africa (Fishpool & Evans 2001). It is also now well established that the Gulf of Guinea islands are of global importance for their levels of endemism in other taxa as well (Jones 1994).

Practical measures to safeguard this biodiversity in São Tomé and Príncipe were initiated by ICBP and IUCN with European Community funding in the late 1980s (Jones & Tye 1988, Harrison & Steele 1989, Fa 1991, Jones *et al* 1991, Atkinson *et al* 1994). Foremost among these measures was a proposal that had been put forward earlier (BDPA 1985) for the legal protection of key forest areas as *zonas ecológicas* on the two islands. Legislation was formally promulgated in 1993 to protect some 245 km^2 in central and southwestern São Tomé and 45 km^2 in southwestern Príncipe (RDSTP 1993). By this time, São Tomé & Príncipe had become part of the European Union-funded ECOFAC project, which was established in 1992 to promote the conservation and rational utilisation of rainforest resources in seven central African states, including Equatorial Guinea, Cameroon, Central African Republic, Gabon, Congo and Zaire (now the Democratic Republic of Congo). The ECOFAC project has overseen the establishment of a headquarters and field-station for the Parque Natural Ôbo, defined the limits of the national park boundaries for both São Tomé and Príncipe, the training of forest guards, and the initiation of various education programmes in conservation and rural development. In addition, ECOFAC has included various coastal zone and marine conservation issues within its remit (GGCG 1996d), such that the boundaries of the Parque Natural Ôbo de São Tomé also include the small but important area of mangrove on the Rio Malanza in the south and the dry forest near Lagoa Azul in the north, and additionally afford some protection to important breeding sites of marine turtles. ECOFAC also sponsored the publication of a preliminary Red Data List of threatened animals (Gascoigne 1995), the first illustrated field guide to the birds of São Tomé and Príncipe (Christy & Clarke 1998), and a guide to orchids (Stévert & Oliveira 2000). Further information about these initiatives can be found in the ECOFAC project's newsletter *Canopée* and on its website at <http://www.ecofac.org>.

In Equatorial Guinea, ECOFAC was primarily concerned with the continental reserve of Monte Alén, not with the islands of Bioko or Annobón. Conservation projects in Bioko have instead been initiated by the Proyecto de Conservación y Ecodesarollo en el Sur de la Isla de Bioko, supported by the Spanish Asociación Amigos de Doñana (GGCG 1996b). Annobón remains relatively neglected and, although listed as an Area Protegida under Law 8/1988 of 31 December 1988 which regulates wildlife, hunting and protected areas, publication of the corresponding development regulations and plans for use and management are still awaited (República de Guínea Ecuatorial 1998).

There is increasing interest in the biodiversity and conservation of the Gulf of Guinea outside the region itself. A meeting at Jersey Zoo in 1993, sponsored by the Jersey Wildlife Preservation Trust, first brought together biologists working on the Gulf of Guinea islands and resulted in the publication of a symposium volume *Biodiversity Conservation in the Gulf of Guinea Islands* (Fa & Juste 1994), as well as the formation of an informal association, the Gulf of Guinea Conservation Group. Beginning in 1995, the GGCG kept note of new biological discoveries and issues of conservation concern on the islands, initially through its occasional publication, the *Gulf of Guinea Conservation Newsletter*. The GGCG now acts through an informal grouping of interested parties known as the Gulf of Guinea Islands' Biodiversity Network and its website is at <http://www.ggcg.st>.

SYSTEMATIC LIST

Presentation of species accounts

Scientific and English names
The sequence and nomenclature follow *The Birds of Africa* unless otherwise stated. Our main departure from these authors is that we follow Naurois (1983a) and Collar & Stuart (1985) in according higher taxonomic rank wherever there is debate about the specific distinctiveness of particular taxa (see Amadon & Short 1992, Haffer 1992, Mallett 1995). Additional English names of passerines are taken from Collar & Stuart (1985) or other recent sources. Abbreviated authorities are given for scientific names; the full references for the original descriptions of endemic species and subspecies (but not for non-endemic taxa) are included in the bibliography.

Subspecific names are given in the species account headings for endemic taxa only; non-endemic subspecies are named in the text for other polytypic species. Superspecies are those used by *The Birds of Africa* (non-passerines) and Hall & Moreau (1970) (passerines) unless otherwise stated.

Local names
Following the scientific and English name of each species, local names from the three islands are given where known. Portuguese and Spanish ornithological names are given (indicated by 'Port' and 'Sp' respectively), if found in literature about the islands. These names are not necessarily current on the islands and may not conform to modern Portuguese and Spanish usage. A reference for each vernacular name is given, abbreviated for the commonest sources as follows:

B	Newton (in papers by Bocage, with date indicating which paper)
Bar	Barrena (1911)
Bas	Basilio (1957)
C&C	Christy & Clarke (1998)
F	Frade (1958)
F&V	Frade & Vieira dos Santos (1977)
J&T	ourselves
K	Keulemans (1866)
M	M. Melo (pers comm)
N	Naurois (date indicates which publication)
N&CA	Naurois & Castro Antunes (1973)
R	Reinius (1985)
S	Fea (in papers by Salvadori, with date indicating which paper)

Where a bird occurs on more than one island, we have tried to indicate to which the name refers by the use of the letters P, ST and A. As the spellings of a local name given by different authors vary (often according to the author's nationality), we have usually presented the version closest to that which a native Spanish or Portuguese speaker would use. Of the names for Annobón, those given by Basilio (1957) are probably more reliable and up to date than those of Barrena (1911).

Status code
The status of each species is given in abbreviated form at the head of each account as follows:

BS	Breeding seabird
NBS	Non-breeding seabird

RB   Resident breeder
PM   Palearctic migrant of regular occurrence
VPM   Vagrant Palearctic migrant
AMNB   Afrotropical migrant, not breeding on the islands
VA   Vagrant Afrotropical species
V   Vagrant, geographical origin unknown
E   Extinct
?   Status unclear
–   Not recorded on that island

Three status abbreviations are given for the three islands, in the order Príncipe, São Tomé, Annobón. Sightings at sea are referred to the nearest island, unless far offshore. Species of doubtful occurrence appear in square brackets. Taxa endemic to the islands are marked:

*   endemic subspecies of mainland species
**   endemic species
***   endemic genus

Texts of species accounts

Each species account contains statements of distribution outside (brief) and within the islands, abundance, status, migrations, habitat, behaviour (summarised and referenced when taken from published sources), breeding season and threats. Breeding season data are given where quoted in literature; note that the breeding seasons quoted by Naurois are not reliable: his papers contradict each other on this point, and often contradict the evidence collected by other observers. Taxonomic discussion is included where it is of particular interest or we have carried out new assessments.

Abundance/conspicuousness ratings correspond loosely to those used in *The Birds of Africa*:

Very abundant   >100 may be seen or heard in suitable habitat per day
Abundant   11–100 may be seen or heard in suitable habitat per day
Common   1–10 may be seen or heard in suitable habitat per day
Frequent   often seen but some effort required (not every day)
Uncommon   several records per year
Rare   one record per several years (resident spp)
Vagrant   one record per several years (non-residents)

These ratings are used as adjectives or adverbs (eg frequently seen).

Our own unpublished observations are referred to by 'J&T' (observations in July–August 1987) and 'J&B' (observations by PJJ and J.P. Burlison in July–August 1988). Tape recordings mentioned in the text, made by us or by P.D. Alexander-Marrack, are deposited with the UK National Sound Archive, London.

Locations of specimens

In the Systematic List we have tried to include mention of the most important collections of specimens of endemic taxa but have usually not done so for non-endemics. Museums with substantial collections or important specimens are listed below, with acronyms for those most frequently mentioned in the species accounts:

AMNH – American Museum of Natural History, New York, USA
Contains the collections and field notes of J.G. Correia, used as the basis for Amadon's (1953) review.

BMNH – The Natural History Museum, Tring, UK (formerly British Museum (Natural History))
Contains the collections of Thomson, Alexander, Snow and Fry, plus 19th-century specimens from a variety of collectors and dealers, including some important specimens collected by Newton and Fea, donated by Bocage and Salvadori respectively. Many types are included in this collection.

Bocage Museum (Museu Nacional de História Natural, Departmento de Zoológico e Antropológico (Museu Bocage), Universidade de Lisboa; formerly Museu de Lisboa), Lisbon, Portugal
Contained most of the important Portuguese collections made in the 19th century, including most of the specimens collected by Newton and described by Bocage. Most of the museum's bird specimens were destroyed in two fires, in 1910 and on 18 March 1978.

Coimbra Museum (Museu Zoológico, Universidade de Coimbra), Coimbra, Portugal
Specimens collected by A.F. Moller and others in the 19th century. Includes some albino specimens.

IICT – Centro de Zoologia, Instituto de Investigação Científica Tropical (formerly Junta de Investigações do Ultramar), Lisbon, Portugal
A small collection, including specimens of *Ploceus velatus peixotoi* and the collections of Frade.

MNHN – Museum National d'Histoire Naturelle, Paris, France
Includes some important specimens donated by Bocage, and the collections of Naurois.

Museo di Storia Naturale 'Giacomo Doria', Genoa, Italy
Some specimens collected by Fea, described by Salvadori.

Museo di Zoologia Sistematica, Universitá di Torino, Turin, Italy (formerly Museo di Zoologia e Anatomia Comparativa)
Contains some specimens, including perhaps some of Fea's.

MNCN – Museo Nacional de Ciencias Naturales, Madrid, Spain
A few specimens from the 1959 expedition to Annobón (J. Pérez del Val *in litt*).

Museo Regionale di Scienze Naturali, Turin, Italy
Some specimens collected by Fea, described by Salvadori.

Museum für Naturkunde, Humboldt-Universität, Berlin, Germany
A few of Newton's and Correia's specimens.

Museum Heineanum, Halberstadt, Germany
A few early specimens.

Swedish Museum of Natural History (Naturhistoriska Riksmuseet), Stockholm, Sweden
Small collection, listed by Reinius (1985).

Szczecin Museum (Muzeum Narodowe w Szczecinie), Szczecin, Poland
Collection of Dohrn.

Übersee-Museum, Bremen, Germany
Contained some 19th-century specimens from German collectors, although many of
these were destroyed in the Second World War.

ZMH - Zoologisches Museum Hamburg, Hamburg, Germany
Contained extensive collections, including those of Weiss, which were described by
Hartlaub. The museum was destroyed in the Second World War and most of the
specimens and catalogues were lost, although some survive in the rebuilt museum,
including those of *Halcyon malimbica dryas*.

PROCELLARIIDAE

[Bulwer's Petrel   *Bulweria bulwerii*   ?;?;–
                                 (Jardine & Selby 1828)

Breeds on Atlantic islands, migrates southwards in non-breeding season; occurs along W African coast south to Liberia (Brown *et al* 1982).
     Four reported at sea between São Tomé and Príncipe, Mar 1992 (I. Sinclair *per* P. Christy *in litt*, Christy & Clarke 1998). This record requires confirmation.]

[Cory's Shearwater   *Calonectris diomedea*   ?;?;–
                                 (Scopoli 1769)

Port: Pardela-de-bico-amarelo, Cagarra (C&C).

Breeds on islands in central Atlantic and Mediterranean; migrates as far as South Africa in non-breeding season; common west of the islands (Brown *et al* 1982, Lambert 1988).
C 20 seen flying north between São Tomé and Príncipe, Mar; no details given (I. Sinclair *per* P. Christy *in litt*, Christy & Clarke 1998).]

[Great Shearwater   *Puffinus gravis*   ?;?;–
                                 (O'Reilly 1818)

Port: Pardela-de-bico-preto, Cagarra (C&C).

Breeds S Atlantic islands. Non-breeding visitor throughout Atlantic, including entire W African coast except Gulf of Guinea (Togo to Zaire) (Brown *et al* 1982, Grimes 1987). No certain records.

Príncipe and São Tomé Sighted between the islands by Reinius (1985). This record requires confirmation owing to the possibility of confusion with Cory's Shearwater (S. Reinius *in litt*). Also recorded in late Sep and Oct, without details (Christy & Clarke 1998).]

Sooty Shearwater   *Puffinus griseus*   ?;?;NBS
                                 (Gmelin 1789)

Port: Procelária-pardacenta, Tantónio (C&C).

Breeds on islands in southern oceans, wanders north in non-breeding season in Atlantic, including Gulf of Guinea and Bioko. Only definitely recorded Annobón (sight records doubtful owing to possibility of confusion with Bulwer's Petrel); no breeding records.

São Tomé One sighted *c* 0°58'S, 7°40'E, southeast of the island, 11 Sep 1979 (Cadée 1981).

Príncipe and São Tomé Reinius (1985) reported sighting it at sea between these islands, but this requires confirmation (S. Reinius *in litt*). Reported Aug–Sep by Christy & Clarke (1998), without details.

Annobón An adult male captured offshore by fishermen (Bocage 1893b, 1903) is apparently the only record; Bocage gave no date.

[Little Shearwater   *Puffinus assimilis*   –;–;–
                                 Gould 1838

Breeds central Atlantic; non-breeding visitor to W African coast south to Sierra Leone, and southern African coast; no records Gulf of Guinea (Brown *et al* 1982).

Included by Christy & Clarke (1998) based on a 1949 observation by the Oxford Expedition passed to Christy by W.R.P. Bourne (P. Christy *in litt*), but Bourne (*in litt*) lists this only as 'Dusky (Little or Audubon's?) Shearwater', ie *P. affinis* or *lherminieri*, so that the species remains unconfirmed.]

HYDROBATIDAE

[Wilson's Storm-Petrel   *Oceanites oceanicus*   –;–;–
                                 (Kuhl 1820)

Port: Casquilho, Ave-das-tormentas, Calcamares (C&C).

Breeds on S Atlantic islands; common all African seas (Brown *et al* 1982), including west of the islands (Lambert 1988).

Reported in Gulf of Guinea Jul–Sep (Christy & Clarke 1998), apparently based on 19 Sep 1949 observations of the Oxford Expedition passed to Christy by W.R.P. Bourne (P. Christy *in litt*); the locality was 'ca 2°°E [*sic* = S] 09°E SE of S Tomé' (W.R.P. Bourne *in litt*). This record requires confirmation.]

[Black-bellied Storm-Petrel *Fregetta tropica*   –;–;–
                                 (Gould 1844)

Breeds S Atlantic and Indian Ocean islands, dispersing northwards in non-breeding season; no records Gulf of Guinea (Brown *et al* 1982).

Reported for the Gulf of Guinea in Sep (Christy & Clarke 1998), based on a 1949 observation by the Oxford Expedition passed to Christy by W.R.P. Bourne (P. Christy *in litt*), but Bourne (*in litt*) lists this record as only probable.]

[British Storm-Petrel   *Hydrobates pelagicus*   ?;?;–
                                 (Linnaeus 1758)

Port: Casquilho, Ave-das-tormentas, Calcamares, Alma-de-mestre (C&C).

Breeds N Atlantic, winters to S Atlantic, but not in Gulf of Guinea (Ghana to Angola) according to map in Brown *et al* (1982, see also Grimes 1987). The report by Reichenow (1900–05) for Bioko and São Tomé was doubted by Louette (1981), but it has been reported subsequently around the islands; no breeding records.

Seen at sea along the Guinea rise, on which the islands stand, by Robins (1966) and at sea between São Tomé and Príncipe by Reinius (1985). Reported for Gulf of Guinea in Mar and Nov (Christy & Clarke 1998). Owing to the possibility of confusion with Wilson's and Leach's Storm-Petrels, unsupported sight records must be considered subject to doubt.]

Madeiran Storm-Petrel   *Oceanodroma castro*   **(B?)S;(B?)S;NBS**
(Harcourt 1851)

Port: Caniboto or Camboto (B 1891, 1904).

Breeds on islands in Atlantic and Pacific, non-breeding dispersal in East Atlantic. Recorded around and on the islands but no definite breeding records; sight records perhaps subject to doubt. Specimens from São Tomé suggest that the island may support a distinct subspecies (Harris 1969), which would almost certainly qualify as threatened (Gascoigne 1995).

At Sea Reported at sea along the Guinea rise (on which the islands stand) by Robins (1966) and sighted at sea off São Tomé by Reinius (1985). Seen frequently between São Tomé and Annobón in 1989 (Harrison & Steele 1989).

Príncipe Gascoigne (1995) listed it as breeding on Pedras Tinhosas but presented no evidence. One seen at sea off the Tinhosas, Jul 1997 (Monteiro *et al* 1997).

São Tomé May breed on islets off the coast; feathers, probably of a petrel, were found in holes in the volcanic rock of one of the Sete Pedras islets by R. de Naurois (*per* P.L. Britton *in litt*). One was captured far offshore by Angolares fishermen (Bocage 1891, 1904, Salvadori 1903b). Correia obtained 4, which flew into a lighted room at night in November and December 1928 at Roça Jou in the SW (Correia 1928–29b); one of these had enlarged gonads and another was a juvenile in fresh plumage (Amadon 1953, Christy & Clarke 1998), so the species may breed on the main island. Others reportedly fly into windows of the Pousada Boa Vista on foggy days (L. Mario *per* M. Melo *in litt*). Seen in Mar and Jul 1997 at sea 8–19 km east of Santana; reported by fishermen as common all year at sea off São Tomé, and especially during the dry season (Monteiro *et al* 1997).

Annobón Barrena (1911) lists a 'black bird a little smaller than the 'Ígualé'' (Brown Noddy *Anous minutus*), and called 'Doló' by the locals. Basilio (1957) was told that it 'had its abode' in the interior of the island but fed at sea. This suggests that a population of the present species or another petrel breeds on Annobón.

Breeding No firm evidence, but one captured at sea E of Santana, Jul 1997, had a fully developed brood patch; however, it could have flown from breeding colonies in the Cape Verde islands (Monteiro *et al* 1997).

[Leach's Storm-Petrel   *Oceanodroma leucorhoa*   –;–;–
(Vieillot 1818)

Casquilho, Ave-das-tormentas, Calcamares, Alma-de-mestre (C&C).

Non-breeding visitor to entire west coast of Africa (Brown *et al* 1982); common west of the islands (Lambert 1988) and reported about 380 km south of São Tomé in Nov 1979 (Cadée 1981).

Reported near São Tomé and/or Príncipe in Dec–Jan, by Christy & Clarke (1998), based on observations by P. Christy (between São Tomé and Príncipe, Dec 1992) and I. Sinclair (two individuals, Mar 1992) (P. Christy *in litt*). The dates do not quite correspond and we consider that these observations require confirmation.]

PHAETHONTIDAE

[Red-billed Tropicbird	*Phaethon aethereus* Linnaeus 1758	**NBS?;NBS?;–**

Port: Rabo-de-junco (C&C).

Tropical oceans; *P. a. aethereus* breeding in Atlantic on Ascension Is., Cape Verde islands and Iles Madeleines.

Although the map in Brown *et al* (1982) shows this species occurring around the islands, the only evidence that it does so is more recent: one seen by I. Sinclair around Sete Pedras, Mar 1992, and one around Pedras Tinhosas, Jul 1996 (Christy & Clarke 1998, P. Christy *in litt*). We consider that further evidence is required before accepting this species as confirmed.]

White-tailed Tropicbird	*Phaethon lepturus* Daudin 1802	**BS;BS;BS**

STP: Rabo-de-junco, Coconzuco (K, F), Cocozuco (Monteiro *et al* 1997); Rabo-de-palha-de-cauda-branca (Port: N 1994). A: Aguedéguedé (B 1893b, Bas); Rabijunco (Bas).

Oceanic, pantropical; in Atlantic *P. l. ascensionensis*. Breeds on all three islands; also recorded at sea between the islands (Basilio 1957).

Príncipe Seen about the coast by Dohrn (1866) and Keulemans (1866), and over lowland plantations (Atkinson *et al* 1994). In only small numbers around Pedras Tinhosas (Naurois 1973a); at least 10 around Tinhosas and Pedra da Galé, Oct 1991 (Sargeant 1994). No proof of breeding on the main island, but suspected of breeding in substantial numbers on Ilhéu Boné do Jóquei (Naurois 1973a); *c* 10 pairs bred on Ilhéu dos Mosteiros (NE of Príncipe) in the early 1970s and 1997 (Naurois 1973a, 1994, Monteiro *et al* 1997); one bird was seen in a rock crevice on Pedras Tinhosas by R. de Naurois (*per* P.L. Britton *in litt*), and perhaps an irregular breeder there (Naurois 1973a); also breeding on Ilhéus Bonézinho and Tinhosa Grande (Monteiro *et al* 1997).

São Tomé On the main island, found on coastal and inland cliffs, often far into the interior, as at Monte Café and Cão Grande; common on W and E coasts and commonly seen flying over northern savannas (Jones & Tye 1988). Formerly abundant breeder on Ilhéu das Cabras (Bocage 1889c), where 30–40 pairs 'lodged' in the 1960s and 70s (Naurois 1973a); none on land there Dec 1996, Mar and Jun 1997 (Monteiro *et al* 1997); it is not clear whether it still breeds there. Also recorded breeding on Sete Pedras, Ilhéu das Rolas, Ilhéu de Santana and Ilhéu Quixibá (Bocage 1889c, 1891, Bannerman 1915a, Frade 1958, Naurois 1973a, Nadler 1993, Monteiro *et al* 1997), and on the main island on trees (Naurois 1973a, 1994), on cliffs on the NW coast and by Ribeira Palma and in the valleys of rivers Abade, Ió Grande and Contador (Jones & Tye 1988, Monteiro *et al* 1997, see also Günther & Feiler 1985).

Immatures are caught by local people for food (Jones & Tye 1988), but most nests are in such inaccessible places that the population cannot be endangered by cropping at present levels.

Annobón In 1892 Newton found it breeding in rock crevices at 500 m on 'Pico Estephania' (Bocage 1893b, the peak – Pico Estephania is not found on modern maps). Noted as breeding in 1911 by Schultze (1913). In the 1950s about 50 pairs bred on the main island (Fry 1961), probably including Montaña de Capuchinos (NE) (Basilio 1957). Seen on steep inland cliffs in the 1960s (Robins 1966), and probably still common (Harrison 1990).

Habits Feeds far out at sea on fish and squid; non-incubating partner frequently circles and hovers before nest; fairly silent (Naurois 1973a).

Breeding The egg is described by Bocage (1889c, 1891), the egg and nest-sites by Naurois (1973a). Mainly breeds on rock ledges and in crevices, but nest-sites also include tunnels in thick grass, and tall trees (Basilio 1957, Nadler 1993, Naurois 1994), inland up to 500 m altitude on São Tomé (Naurois 1973a). One male collected on Annobón, 30 Sep, had enlarged gonads (Basilio 1957). On Ilhéu das Cabras, presence of juveniles indicated egg-laying about Oct–Nov (Frade 1958), and birds were visiting the islet in Sep (Bannerman 1930–51 vol. 8). On Ilhéu de Santana, a young bird in a nest tunnel was near fledging, Aug (Nadler 1993). On Sete Pedras, a female taken on the nest in Oct had just laid (Frade & Vieira dos Santos 1977). Birds were incubating Mar and Jul on Sete Pedras, Jun on Mosteiros and Jul on Rolas and Tinhosa Grande; downy chicks in Jan on Sete Pedras, Jun on Mosteiros; fledgling Apr on Bonézinho (Monteiro *et al* 1997). Apparently breeding Jul–Aug 1987 on mainland São Tomé (Jones & Tye 1988). Naurois (1973a) considered that individuals bred at different times of year, so that chicks or eggs could be found year-round; the reproductive cycle of individuals (annual or longer) is unknown.

SULIDAE

[Cape Gannet *Sula capensis* ?;?;–
(Lichtenstein 1823)
Port: Alcatraz do Cabo (C&C).

Breeds southern African coasts, young reportedly move north as far as Gulf of Guinea. Reported, without details, from Bioko (Pérez del Val *et al* 1994).

Although the map in Brown *et al* (1982) shows this species occurring around the islands, there is no firm evidence that it does so, although an immature of this species or Northern Gannet *S. bassana* was reported at Praia das Conchas, Aug 1991 (Sargeant 1994), while another (or the same?) immature seen by I. Sinclair off northern São Tomé in Aug 1991 has been referred to this species (Christy & Clarke 1998, P. Christy *in litt*).]

Masked Booby *Sula dactylatra* V?;?;V?
Lesson 1831

Oceanic, pantropical; in Atlantic *S. d. dactylatra*. Rarely found in E. Atlantic (Brown *et al* 1982). Vagrant Príncipe and Annobón.

Príncipe One specimen, a male taken at Pedras Tinhosas in 1960 (Frade & Vieira dos Santos 1977); one adult seen on Tinhosa Grande, Jul 1996 (Christy & Clarke 1998). Gascoigne (1995) listed it as breeding there but presented no evidence.

Six seen at sea between São Tomé and Príncipe, Mar 1992 (Christy & Clarke 1998).

Annobón One record, an adult seen on Santarém islet, May 1965 (Robins 1966).

[Red-footed Booby   *Sula sula*   ?;–;–
                               (Linnaeus 1766)

Pantropical but only vagrant W Africa (Brown *et al* 1982).

Príncipe At least one brown immature reported by P. Christy (*in litt*) around Pedras Tinhosas, Jan 1995; one reported by I. Sinclair at Ilhéu Bom-bom, Apr 1995 (Christy & Clarke 1998, P. Christy *in litt*). These records require confirmation, owing to the possibility of confusion with Brown Booby.]

Brown Booby   *Sula leucogaster*   **BS;BS;NBS**
                               (Boddaert 1783)

STP: Muinbini, Muinbinhi (K); Matchia-vagé (B 1891, 1904); Malvaxi, Pato-marinho (Monteiro *et al* 1997); Alcatraz, Ganso-patola-pardo (Port: F&V, N 1994); Bobo (R). A: Hoho (B 1903); Ioj (Bar); Joj (Bas: pron. 'Hoh'); Alcatraz de vientre blanco (Sp: Bas).

Oceanic, pantropical; *S. l. leucogaster* in E. Atlantic, Senegal to Gulf of Guinea (Brown *et al* 1982). Breeds Príncipe and São Tomé but not, apparently, on Annobón.

Príncipe Dohrn (1866), Keulemans (1866) and Harrison & Steele (1989) found them common on rocks off the S and W coasts; Keulemans thought that they fished mainly in the morning, and rested from midday onward. Breeds on several offshore islets: on Pedras Tinhosas, many hundreds of breeding pairs found by Naurois (1973a) but only *c* 10 pairs mentioned for Tinhosa Grande by Naurois (1994); several thousand birds found on Tinhosas by Christy (1995b); breeds also on Pedra da Galé (Frade 1958, 1959) although Naurois (1973a) thought that weather conditions would rarely allow any nest on the tiny, wave-washed Galé to reach full term; also breeds, perhaps irregularly (<10 pairs), on Ilhéu dos Mosteiros (Naurois 1973a, 1994); suspected of breeding on Ilhéu Boné do Jóquei by Naurois (1973a) and young found there by Christy (1995b). Recently, fishermen from Príncipe and São Tomé have begun collecting chicks on Tinhosa Grande, for sale on São Tomé, which may threaten that important population with extinction (Gascoigne 1995, Christy 1995b).

São Tomé Probably first recorded about 1500 by G. Piriz ('Alcatrazes muytos': Monod *et al* 1951). Breeds on Sete Pedras islets (Bocage 1891, 1904, Nadler 1993). Frequently seen fishing inshore (Günther & Feiler 1985, Eccles 1988, Jones & Tye 1988, Harrison & Steele 1989).

Annobón Common at sea and on offshore islets for most of the year; said not to breed there, but to depart at the beginning of the wet season (Nov or early Dec) to breed 'on Príncipe' (Basilio 1957), although São Tomé is closer. Only one seen during a one-day visit in Feb 1989, at a time when it was also seen only rarely off São Tomé and Príncipe (Harrison 1990). Two males and 2 females were collected by Newton on Tortuga islet, where he believed it was breeding (Bocage 1893b); 10–12 birds there in May (1964 or 1965) with others flying nearby (Robins 1966). Each of the four southern islets (Adams, Santarém, Escobár, Fernando Póo) had 'colonies' of 100s in May (Robins 1966), presumably breeding. Harrison (1990) believed that it had declined at Annobón since the 1960s, but frequency of sightings may depend heavily on the time of year.

Breeding Keulemans (1866) gave a description obtained from local people of the egg and nest; Naurois (1973a, 1994), Monteiro *et al* (1997) and Christy & Clarke (1998) described nests, eggs and nestlings. An egg of this species described by Bocage (1891) as collected at Martim Vaz, was probably not from the islands (no such place appears on any map or gazetteer consulted), but perhaps came from the islet of Martin Vaz, or a neighbouring islet, off Brazil.

Keulemans (1866) stated that Príncipe birds bred in Dec and Jan. On Ilhas Tinhosas, Christy (1995b) reported one downy young and several juveniles in brown plumage in Jan, and more numerous downy young in Aug 1995, while Christy & Clarke (1998) reported adults incubating eggs, and no chicks, in Jan 1996; on Tinhosa Grande, Monteiro *et al* (1997) reported incubating adults and downy young Apr, Jun–Jul. About 200 pairs breeding on Sete Pedras, many with nestlings, Apr 1991 (Nadler 1993); Monteiro *et al* (1997) found downy young there and on Bonézinho in Jul. Nestlings were collected on Pedra da Galé in Nov (Frade 1958, 1959). One bird collected on the southern islets of Annobón, sometime between Aug and Nov, was coming into breeding condition (Basilio 1957). The above data suggest laying occurs at least Jan–Jun (from incubation and nestling periods given by Brown *et al* 1982); Naurois (1973a) gave laying period as Nov–May, with most nests on Tinhosa Grande containing eggs and young chicks in Jan, and only a few well-grown chicks found there during Aug; he thought that the reproductive cycle was annual. Christy & Clarke (1998) suggest that there are two breeding peaks per year, in Jan and Jul–Aug.

PHALACROCORACIDAE

[White-breasted Cormorant   *Phalacrocorax carbo*   ?;?;–
(Linnaeus 1758)

Near cosmopolitan; NE Atlantic south to Sierra Leone and SE Atlantic from Angola southwards (*P. c. lucidus*). Unconfirmed sightings Príncipe and São Tomé.

Príncipe Naurois (1987b) reported a possible sighting of 'two or three' immatures in poor viewing conditions.

São Tomé One reportedly seen on SE coast at Ribeira Peixe, 14 June 1983 (Günther & Feiler 1985). The bird was flying over the sea. Given the difficulties of discriminating this species at a distance from Reed Cormorant and the distance of São Tomé from its usual range, we consider that this record requires confirmation. However, a local hunter also claimed to have seen large cormorants on the E coast between Rio Ió Grande and Santa Cruz previously (Günther & Feiler 1985).]

Reed Cormorant   *Phalacrocorax africanus*   –;RB;–
(Gmelin 1789)

Pata d'aua (B 1891, 1904); Corvo marinho (F, N 1972b); Pato-marinho (C&C). Port: Corvo-marinho-africano (C&C).

Afrotropical including Bioko (*P. a. africanus*). Resident on São Tomé; records on Príncipe were implied by Naurois (1994) but there is no documentary evidence for its occurrence there.

First collected by Weiss, although perhaps noted earlier by Lopes de Lima, and found by most ornithologists since (Naurois 1987b). At present, common along rivers and on lagoons in W and SE São Tomé and frequently seen flying over the forest (Jones & Tye 1988, Christy & Clarke 1998). Also recorded from SW, N, and Ilhéus Quixibá and das Rolas (Bocage 1891, 1904, Frade 1958, Eccles 1988, Nadler 1993); may breed on Ilhéu Pato Bravo (Christy & Clarke 1998). Inhabits fast mountain streams as well as slower, broader, lower reaches; also regularly fishes on the sea (Jones & Tye 1988).

Breeding In breeding condition Nov and Dec (Amadon 1953); one collected Nov with egg ready to lay (Correia 1928–29a); nests with eggs laid Oct–Dec, Feb, in mangroves at Diogo Nunes and at Lagoa de Malanza (Frade 1958, 1959, Naurois 1972b, 1987b, 1994); juveniles seen at Ponta da Agulha (SE), Oct (Frade 1958) and at Rio Ió Grande, Apr and Jun (Günther & Feiler 1985, Eccles 1988). Nests were described by Naurois (1987b, 1994). Frade (1958) reported juveniles and nesting in association with Cattle Egrets and Reef Herons.

[FREGATIDAE]

[Ascension Frigatebird　　　*Fregata aquila*　　　　　　　**?;–;–**
　　　　　　　　　　　　　　(Linnaeus 1758)

Port: Rabil, Fragata (C&C).

Ascension Island and south tropical Atlantic.

　　Although the map in Brown *et al* (1982) shows this species occurring around the islands, there is no firm evidence that it does so (early reports from the Gulf of Guinea date from before more than one species of *Fregata* was recognised: W.R.P. Bourne *in litt*). More recently, at least 4 individuals of a *Fregata* sp were seen over Pedras Tinhosas, Aug 1991 (Sargeant 1994, I. Sinclair *per* P. Christy *in litt*), and 2 immatures reported by I. Sinclair between São Tomé and Príncipe in Mar 1992 (Christy & Clarke 1998, P. Christy *in litt*). These records are perhaps more likely to have been Magnificent Frigatebird *F. magnificens*, which is less marine, more a coastal bird, and may wander from W Atlantic (W.R.P. Bourne *in litt*).]

ARDEIDAE

Little Bittern　　　　　　*Ixobrychus minutus*　　　　　**(R?)B;–;–**
　　　　　　　　　　　　　　(Linnaeus 1766)

Port: Garcenho-pequeno-africano (N 1994, C&C).

Afrotropical resident (*I. m. payesii*) and Palearctic migrant. One breeding record, Príncipe.

　　The only record was by Naurois (1972b, 1983a, 1987b, 1994) who reported an immature collected on a nest in a bush on an islet in a pond in a large marshy valley south of Santo António, in Jan 1971 (Naurois 1987b; some of these facts were misquoted by Christy & Clarke 1998). The nest contained broken shell, having perhaps been used the previous Sep.

[Black-crowned Night Heron　　*Nycticorax nycticorax*　　　**–;?;–**
　　　　　　　　　　　　　　(Linnaeus 1758)

Near-cosmopolitan; in Africa *N. n. nycticorax*. One record São Tomé.

　　Harrison & Steele (1989) reported a sighting (second-hand) of a night-heron near an inland river.]

[Squacco Heron   *Ardeola ralloides*   ?;–;–
                                       (Scopoli 1769)
Port: Garça-caranguejeira, Papa-ratos (C&C).

Resident and Palearctic winter visitor in sub-Saharan Africa (Brown *et al* 1982).

One immature seen by P. Christy in marshes on the banks of the Rio Papagaio at Santo António, Jan 1996 (Christy & Clarke 1998, P. Christy *in litt*). Given the slight possibility of the occurrence of other *Ardeola* vagrants, we regard this record as provisional.]

Cattle Egret   *Bubulcus ibis*   **(R?)B;(R?)B;AMNB**
                                       (Linnaeus 1758)

STP: Gaça, Garça (B 1891, R); Garça-boieira, Carraceira (Port: F&V, N 1994, C&C). A: Ngasa babanc, Garcilla bueyera (Bas).

Near-pantropical; in Africa *B. i. ibis*. Breeds and probably resident on São Tomé and Príncipe, non-breeding visitor to Annobón.

São Tomé and Príncipe First recorded on Príncipe in 1954 (Frade 1958) and perhaps a recent arrival there, although known on São Tomé since at least the mid-19th century (Hartlaub 1850). Now common to abundant on both islands, singly or in flocks up to 80 (Günther & Feiler 1985) in open areas, including roadsides, clearings, gardens, open plantations, along the coast, and in the northern savannas of São Tomé, where attracted to grass fires (Jones & Tye 1988). Occurs up to 1200 m on São Tomé (Snow 1950, Sargeant 1994). Large breeding colony and non-breeding roost in mangroves at Água Izé (Christy & Clarke 1998). Recorded breeding on Ilhéu das Rolas (Bocage 1891), on Ilhéu das Cabras (Correia 1928–29b), near the airport in Aug 1960 (Basilio 1963), and more recently in mangroves at Praia das Conchas, Santana and near the airport (*c* 15 pairs, 50 pairs and 5 pairs respectively: Nadler 1993). One bird recently found on Ilhéu das Rolas (Nadler 1993) but not clear whether still breeds there.

Annobón Reports from local people that small white herons arrived before Christmas and stayed 'a couple of months', foraging among the sheep, probably referred to this species (Basilio 1957). Fry (1961) saw one in Jul 1959, and thought it an irregular visitor; it reportedly followed ships to the island from São Tomé, and was commoner in the wet season (Oct–May), with individuals generally staying one or two weeks. However, an arrival about Dec would coincide with its southward migration to Bioko and Mbini (Equatorial Guinea) (Basilio 1957, Brown *et al* 1982).

Habits Appears to eat crabs (Naurois 1987b).

Breeding Sometimes breeds in colonies with other herons or Reed Cormorant (Nadler 1993, Naurois 1987b, 1994), which are pillaged by local people (Naurois 1972b). Nest-sites, nests and eggs described by Naurois (1987b, 1994), eggs from Ilhéu das Rolas by Bocage (1891). In non-breeding plumage, Apr (Eccles 1988); in buff breeding plumage on São Tomé, Jun–Sep (Jones & Tye 1988, Atkinson *et al* 1994); specimens collected Dec had breeding plumage and large gonads (Amadon 1953). Correia (1928–29b) found eggs on Ilhéu das Cabras in early August; Basilio (1963) found many nests in a patch of woodland behind the airport called Diego Nuñes, Aug 1960; Naurois (1987b) found eggs laid Nov–Apr and Jun–Jul; Frade (1958) found nests with young at Ponta da Agulha (SE São Tomé) in Oct and in mangroves at Diogo Nunes (NE São Tomé) in Nov. These records mostly correspond with the breeding season of the southern African population (*cf* Amadon 1953). One individual seen on Annobón in

Jul was also acquiring breeding plumage (Fry 1961). However, Günther & Feiler (1985) found a breeding colony in a baobab *Adansonia digitata* at Praia das Conchas (N São Tomé) in early June. Although the colony contained about 20 nests, some with eggs, others with young, Günther & Feiler (1985) saw no adult in breeding dress and suggested (as did Amadon 1953) that the breeding period was not well defined. Further, Naurois (1972b) reported eggs Dec–Mar, and Nadler (1993) found nestlings 2–3 weeks old in Mar–Apr at Santana (E São Tomé). Naurois (1994) thought that it breeds nearly year-round, with peak Nov–Dec, although he presented no additional evidence for this.

Green-backed Heron   *Butorides striatus*   **RB;RB;VA**
(Linnaeus 1758)

STP: Gallo-d'água (K); Tchongo, Tjonzo, Tchomjo, Tchonze (B 1879, 1888a, 1891, N 1994, C&C); Chuchu (F&V); Garça-de-cabeça-negra (Port: N 1994, C&C).

Pantropical; in Africa *B. s. atricapillus*. Resident São Tomé and Príncipe, vagrant Annobón.

São Tomé and Príncipe First collected by Weiss on São Tomé (Hartlaub 1852) and by Dohrn (1866) and Keulemans (1866) on Príncipe. Reported by most ornithologists since (Naurois 1987b). At present, frequent to common on lower reaches of rivers and occasionally on rocky coasts, and on Ilhéu das Rolas (Bocage 1891, 1904, Jones & Tye 1988, Nadler 1993). Recorded on Príncipe along the coast and on streams up to 450 m (Keulemans 1866, Sargeant 1994).

Annobón One taken by Newton in 1892 at the mouth of the S. Juan river (Bocage 1893b, 1903) appears to be the only record.

Habits Call as in mainland birds; other habits described by Naurois (1987b).

Breeding Local people told Keulemans that it bred on Príncipe in Nov–Dec. Alexander thought it breeding on São Tomé in Jan–Feb (Bannerman 1915a); a juvenile collected there, Aug (Salvadori 1903b); another seen with adults, Apr (Eccles 1988); a pair probably breeding in a colony of Cattle Egrets and Reef Herons, Mar–Apr (Nadler 1993). Naurois (1972b, 1987b, 1994) stated that laying period was extended, including at least Jun–Jan, possibly to Mar or May, but the data summarised by Naurois (1987b) only confirm laying Sep–Dec. Nest-sites, nest and egg described by Bocage (1891) and Naurois (1972b, 1987b, 1994).

Black Heron   *Egretta ardesiaca*   **–;VA;–**
(Wagler 1827)

Port: Garça-preta, Garça-ardósia (C&C).

Mainland Africa. One record São Tomé. Specimens from Annobón collected by Lowe, which were originally identified as this species (Bannerman 1912) are actually of Reef Heron *E. gularis* (R. Prys-Jones *per* J. Pérez del Val *in litt*).

A single individual photographed umbrella-feeding in shallow pools on Rio Angra Toldo (close to the road-bridge), SE São Tomé, Aug 1987 (Jones & Tye 1988); the bird appeared to be hunting small shrimps, though it also caught a small fish. A local woman reported that there were 5 or 6, present all year, which 'fly up-river'.

Western Reef Heron *Egretta gularis* **RB;RB;RB**
(Bosc 1792)

STP: Garça (K, R); Garça-branca (white form: K); Garça-marinha (F&V); Lavadeira (N 1972b); Garça-negra, Garça-marinha (Port: N 1994, C&C). A: Gars (Bar); Ngas, Garceta gris (Bas); Garceta de garganta blanca (Sp: Bas).

Coastal mainland Africa and Bioko; in W Africa *E. g. gularis*. Resident on all three islands.

São Tomé and Príncipe First collected by Weiss (Hartlaub 1852) on São Tomé and Keulemans (1866) and Dohrn (1866) on Príncipe, and reported by most ornithologists since (Naurois 1987b). At present, common on rocky coasts and most islets, and frequent on lower reaches of larger rivers (Jones & Tye 1988), mostly seen singly, occasionally in groups of 2–4, once 24 together (Nadler 1993).

Annobón Apart from the mention by Barrena (1911), which was not accompanied by a description, first recorded in Feb 1909 by Alexander, at the crater lake (Fry 1961). Common in 1955, with groups of 4–5 at the beach at Áquequel (NW), and up to 25 at the lake (Basilio 1957). A colony of 50 bred at the lake in 1959 (Fry 1961) but only 3 seen there in 1989 by Harrison (1990), who speculated that increasing human disturbance may have caused its decline. Apparently colonised Annobón early in the 20th century, as this conspicuous bird was not recorded by Newton or Fea (Fry 1961).

Morphology The population on São Tomé was reported by Amadon (1953) to contain a larger proportion (10 out of 20 specimens collected by Correia) of white-phase birds than the mainland, where white individuals are rare. Other counts on São Tomé gave *c* 20% white (5 or 6 out of 27 or 28 birds, Fry 1961); *c* 30% (7 out of 22 birds, Jones & Tye 1988); *c* 60% (4 out of 7, Eccles 1988); *c* 45% (80 out of 176, Nadler 1993). These counts gave an overall estimate of about one-third white phase, while other reports quoted, without figures, 40% (Atkinson *et al* 1994) or 60–70% (Günther & Feiler 1985, Sargeant 1994) white phase. In contrast, almost all Annobón birds seem to be dark phase (Basilio 1957, Fry 1961 and specimens in BMNH), only one white bird having been reported (Harrison 1990). The limited information from Príncipe includes 6 dark specimens (Amadon 1953 and BMNH) and 2 dark birds seen (J&T), and the statement by Frade (1959) that the white phase was not seen there. This suggests that white-phase birds are rare or absent on Príncipe too.

On São Tomé, in addition to the unusually high proportion of white-phase birds, most or all of the black-phase birds have a white area around the carpal joint (greater coverts and sometimes remiges) (Keulemans 1866, Bannerman 1930–51 vol. 1, Jones & Tye 1988), a plumage type rarely found on the mainland (Amadon 1953). Similarly, on Annobón, many individuals have white areas on the wing, especially on the wing-coverts (Basilio 1957, Fry 1961, Harrison 1990, BMNH specimens). Further description of these and subadult plumages was given by Christy & Clarke (1998).

Habits Stomach contents of an adult male, collected on the crater lake on Annobón, consisted entirely of the fish *Gambusia* (Fry 1961); 2 birds collected on Annobón had stomachs full of small fish (Basilio 1957). *Gambusia* was introduced in about 1945 to combat mosquitoes breeding in the lake (Basilio 1957). V. Fernandes reported that one kind of fish ('muytos peixes enxarrocos') was abundant in the lake in the early 16th century (Monod *et al* 1951). Birds fed on the lake by walking across the *Nitella*-matted surface into which they sometimes sank, when they swam (Fry 1961). However, there was no sign of *Nitella* on the lake in February 1989 (M.J.S. Harrison pers comm). On all 3 islands the species also feeds along the seashore, often in groups (Keulemans 1866, Fry 1961, Jones & Tye 1988).

Breeding Breeds on all three main islands and their associated islets. One stick nest was placed 3 m up a tree (Amadon 1953); reportedly nests in tall trees in the interior of Annobón (Basilio 1957); other nest-sites described by Keulemans (1866); colonies often in mangroves, and often shared with Cattle Egret and Reed Cormorant (Christy & Clarke 1998). The egg was described by Bocage (1891) and young by Keulemans (1866). Of 25 nests examined by Mrs Correia, 23 had one egg, the others 2, but the normal clutch may be larger (Amadon 1953). On Príncipe, Keulemans (1866) thought that they bred Jan–Mar and Dohrn (1866) Mar–Apr. On São Tomé, found breeding by Mrs Correia in Aug (Amadon 1953), by Frade (1958) in Oct–Nov. Nadler (1993) found some 5–10 pairs at Santana and one pair at Praia das Conchas, breeding in mixed colonies with Cattle Egrets *Bubulcus ibis* (*qv*) in Mar and/or Apr. Naurois (1972b) thought they bred Sep–Feb, possibly to Jun, on São Tomé and Príncipe; on both islands, breeding plumage acquired Dec–Apr (Christy & Clarke 1998). On Annobón a female collected on 7 Aug contained a developing egg (Fry 1961).

[Little Egret   *Egretta garzetta*   ?;?;–
                                     (Linnaeus 1766)

Port: Egreta-pequena, Garça-ribeirinha (C&C).

Afrotropical resident and Palearctic migrant (*E. g. garzetta*). Almost certainly occurs on São Tomé and Príncipe, but status there unknown.

Príncipe One observation of a bird on the mudflat at the mouth of the Rio Papagaio, Jan (Christy & Clarke 1998).

São Tomé Moller reportedly collected a specimen on São Tomé in 1885 (Bocage 1887c), though Salvadori (1903b) expressed doubt as to its identity; the specimen should still be in the collection of the Coimbra Museum. Also reported by Snow (1950). However, despite earlier references to the existence of a white phase of Reef Heron *E. gularis* (Keulemans 1866, Bocage 1891, 1904, Salvadori 1903b, Bannerman 1915a), Bannerman (1930–51 vol. 1) did not describe the white phase and recommended identification of *E. garzetta* on the basis of white plumage. Snow's identifications were based on Bannerman (1930–51 vol. 1), so he would not have distinguished *E. garzetta* from white-phase *E. gularis*: there is thus no evidence that the white egrets recorded by Snow (1950) were other than white-phase *E. gularis* (D.W. Snow pers comm). Neither Moller nor Snow recorded white *E. gularis* (see also Frade 1958). More recently, Eccles (1988) reported 6 at various places along the coast, and Christy & Clarke (1998) reported it, without full details, as occurring in small numbers between Dec and Mar, at Pantufo and Praia dos Tamarindos on the north coast.]

[Great White Egret   *Ardea alba*   –;?;–
                                       (Linnaeus 1758)

Port: Egreta-grande, Garça-branca (C&C).

Near-cosmopolitan; Afrotropical resident (*A. a. melanorhynchos*) and Palearctic migrant (*A. a. alba*). Sight records São Tomé.

Two birds reported in 1987, at Monte Café and Praia Micoló (Eccles 1988); the author was convinced of the identifications (S.D. Eccles *in litt*) but, owing to the difficulty of telling this species from congeners, we list it as requiring confirmation.]

| **Purple Heron** | *Ardea purpurea*
Linnaeus 1766 | –;V;– |

Garça-purpúrea, Garça-vermelha (Port: F&V, C&C).

Africa and Eurasia; Afrotropical resident and Palearctic migrant (*A. p. purpurea*). Vagrant São Tomé.

One specimen collected 10 Mar 1970 at Rio Malanza, extreme S (Frade & Vieira dos Santos 1977). One seen at Praia Melão (NE), Dec 1984 (Reinius 1985 and *in litt*).

| **Grey Heron** | *Ardea cinerea*
Linnaeus 1758 | V;V;– |

Garça real, Garça cinzenta (Port: F&V).

Africa and Eurasia; Afrotropical resident and Palearctic migrant (*A. c. cinerea*). Vagrant São Tomé and Príncipe, origin unknown.

Príncipe One seen at Santo António (Reinius 1985 and *in litt*).

São Tomé One specimen, collected 18 Dec 1954 (Frade & Vieira dos Santos 1977).

CICONIIDAE

| **Yellow-billed Stork** | *Mycteria ibis*
(Linnaeus 1766) | –;VA;– |

Port: Flamengo, Falso-flamingo (C&C).

Afrotropical resident. Vagrant São Tomé.

One immature was seen between Santa Catarina and Lembá on the W coast on 6 August 1988 (J&B). This is the only record for the islands.

| **White Stork** | *Ciconia ciconia*
(Linnaeus 1758) | –;VPM;– |

Port: Cegonha-branca (C&C).

Palearctic migrant to tropical Africa (*C. c. ciconia*). Three records from São Tomé.

A specimen collected by Weiss in 1847 (described in the addenda to Hartlaub 1857) was apparently deposited in ZMH (Bocage 1904). Unfortunately, both the specimen and the museum catalogues were destroyed in the Second World War (C. Hinkelmann *in litt*). A second individual stayed on the island for several weeks in 1971, according to Christy & Clarke (1998). The third was photographed at Praia Lagarta, 17 Mar 1999, after heavy storms in the region (A. Gascoigne *in litt*).

THRESKIORNITHIDAE

| **Olive Ibis** | *Bostrychia olivacea rothschildi*
(Bannerman 1919) | RB*;–;– |

P: Corvão (Dohrn 1866, K, F, C&C); Singanga do Príncipe (N 1994); Diógo (C&C).
Port: Galinhola (N 1994); Ibis do Príncipe (C&C).

Subspecies endemic to Príncipe; other subspecies on mainland Africa. Forms a superspecies with *B. bocagei*.

Not seen by an ornithologist during a 90-year period since 1901, when a male was collected at Infante Dom Henrique (SE) by Fea, who reported it rare, living in the forests far from the sea (Salvadori 1903a). Newton did not find it (Bocage 1903) and, *pace* Frade (1958), neither did Correia (*cf* Correia 1928–29b). Previously recorded by Dohrn (1866), who collected a female (Naurois 1973b), and by Keulemans (1866) but only in inaccessible, rocky forests of the mountainous S and W, although both authors also reported that it sometimes descended to the vicinity of Santo António. Keulemans reported it rare, while Dohrn 'saw them daily at great distances' in the southern forests: statements that are not necessarily irreconcilable. These authors suggested that it was conspicuous, noisy and well known to the local inhabitants, who hunted it for food and used its wing-bones for making pipes (Keulemans 1866). The southern forests were reported to have been destroyed or badly damaged in 1906 during tsetse fly control (Barns 1928), although archival circumstantial evidence suggests that forest damage was very limited (G. Clarence-Smith, pers comm).

However, Naurois (1973b) reported a sighting in 1971 by D. Lopes (Naurois 1994 stated that it was in 1973 by D. Nunes, while Naurois & Castro Antunes 1973 reported the last sighting as 1969, perhaps a misprint?) of a bird fitting its description, in a damp ravine 2–3 km S of Santo António, and speculated that it could survive in the now rarely visited, high-altitude forests of the southern half of the island. More recently, Sargeant (1994) reported 2 birds flying over forest 4–5 km from Santo António, in Aug 1991. Seen again by P. Kaestner (*in litt* to P. Atkinson) in Sep–Oct 1997 in forest below the Santo António radio tower. Considering its alleged former conspicuousness, and the restricted area of suitable habitat, it seems likely that the subspecies is now very rare.

Habits Early behavioural and ecological information was provided by Dohrn (1866) and Keulemans (1866): the following statements are from Keulemans unless otherwise noted. It roosted near midday in trees but fed on the ground, taking worms, grubs, snails, large insects, lizards and typhlopid snakes (Dohrn, Keulemans) on or out of the ground. It was difficult to find except when flying, because when disturbed it hid among rocks or fallen trees and when roosting, sat still amongst foliage. Its call was like that of a Raven *Corvus corax* (Dohrn, Keulemans) and was given only in flight. Often calls around dawn and dusk (Christy & Clarke 1998). It was sometimes seen flying with herons or Grey Parrots *Psittacus erithacus*, but more usually alone or in pairs. Islanders thought that its appearance over Santo António presaged disease or death for the white inhabitants (Dohrn) or of an eminent resident.

Breeding Probably nests in trees, but no other information available.

Dwarf Ibis	*Bostrychia bocagei*	–;RB**;–
	Chapin 1923	

Galinha-do-mato, Galinhola (B 1891, 1904, Correia 1928–29b); Singanga de São Tomé (N 1994). Port: Galinhola de São Tomé (N 1994); Ibis de São Tomé (C&C).

Endemic to São Tomé. Forms a superspecies with *B. olivacea* (treated as conspecific by some: see Table 2).

Discovered by Newton, at 6 localities in the 1880s and 1890s: Angolares (E), Morro Gentio (E), Budo-tap'ana (an uncertain locality, said by Naurois (1973b) to be 'près de la Roça Uba Budo, alt. 585 m' but this roça (in the E) is at 243m and there is nowhere over 400 m altitude within 3km), São Miguel (SW), 'Florestas' (ie forests, probably near São Miguel: one specimen was listed by Bocage 1889c as 'Florestas de

S. Miguel', see also Collar & Stuart 1985) and Triumpho (= Triunfo, N) (Bocage 1889b, 1904); 2 individuals were collected at São Miguel and 3 at Angolares (Bocage 1889b,c, 1891). Chapin (1923) reported what may be a further specimen (the type), in MNHN, from 'Rio de São Thomé', identified as Água Tomé near Roça Ubabudo (NE) by Collar & Stuart (1985). Themido (1938) mentions another (juvenile), in the Coimbra Museum, collected by Francisco Quintas in 1888. The only subsequent specimen was a female collected in 1928, at 1100 m above Roça Jou (SW) (Correia 1928–29a,b, Amadon 1953, Collar & Stuart 1985). The maximum number of specimens may therefore be 12, not 6 as quoted by Amadon (1953); however the total is more likely 10 (as the Morro Gentio specimen may be one of the 3 collected at Angolares and the Florestas specimen one of the 2 collected at São Miguel: Collar & Stuart 1985). At least 3 of these were probably destroyed in the fires at the Bocage Museum (Naurois 1983a, 1994, Collar & Stuart 1985), and only 2 are known to remain (the type in MNHN, and AMNH). Correia collected reports of the species from two old men, who had seen 2 at Ió Grande about 1910 and 3 at Roça Jou about 1920 (Correia 1928–29a,b). These records suggest that the species was once widespread in forested country at low and moderate altitudes (perhaps including gallery forest: *viz* Triunfo record in the N savanna), though it has probably always been scarce (Correia 1928–29b, Collar & Stuart 1985).

In the 1960s and early 1970s, the bird was sought fruitlessly in the SW and near Lagoa Amélia and Calvário (over 1000 m in N–centre) (Naurois 1973b). However, in 1988, two local hunters informed PJJ that they had seen it rarely in primary forest at Santo António near the now-abandoned Roça Jou, but never in regenerating forest on the site of old plantations. In 1989, 3 birds ascribed to this species were seen on lava beds in the Rio Ió Grande (B. Schätti, J. Haft *in* Atkinson *et al* 1991). The most recent confirmed records are: one individual seen and photographed on a small stream in primary forest along the steep valley side of the Rio Ana Chaves, Aug 1990 (Atkinson *et al* 1991); another seen four times not far away at 200 m in primary forest between the Ana Chaves and the Ió Grande in Jan 1996 by P. Christy (GGCG 1996a); 3 seen and others heard on ridges at 210–300 m, near Rio Xufexufe, in areas where wild pigs had broken up the soil (T. Gullick *in litt*, Sargeant 1994); 2 pairs, an individual and 2 nests seen near the basins of the Rios Ió Grande (200 m asl) and Martim Mendes (100 m asl), May 1997, in an area where hunters reported killing 16 of the birds 6 months previously (S. d'Assis Lima *in litt*). Also reported from near Estação Sousa and Dona Eugénia and between the rivers Umbugú and Martim Mendes in the Ió Grande and Ana Chaves catchment (Christy & Clarke 1998). Accorded status 'Critically Endangered, D1' (population <50 adults) by BirdLife International (2000).

Habits The birds seem to prefer flatter areas of primary forest, with an open understorey, especially where wild pigs have disturbed the soil; the call is a harsh *karh karh karh* or *karh karh* but the bird is apparently quite silent compared with congeners (Sargeant 1994, Christy & Clarke 1998).

Breeding The female collected by Correia in Nov had enlarged gonads (Collar & Stuart 1985).

PHOENICOPTERIDAE

[Greater Flamingo   *Phoenicopterus roseus*   –;?;–
                              Linnaeus 1758

Mediterranean, Afrotropical and S Asia; W Africa south to Liberia, southern Africa from Congo southwards (Brown *et al* 1982). One doubtful identification São Tomé.

Finsch & Hartlaub (1870) mentioned one specimen in the Übersee-Museum Bremen which was said to have been obtained by Weiss in 1847, but they doubted whether the specimen was genuinely of this species. Strangely, Hartlaub did not mention it in his earlier descriptions of Weiss's collection (Hartlaub 1850, 1852, 1857: see Bocage 1904). Salvadori (1903b) suggested that it was actually a Lesser Flamingo. From the description given by Finsch & Hartlaub (1870) it is impossible to determine to which species the specimen belonged. The specimen was in the Übersee-Museum Bremen until World War II, when it was probably destroyed; it is no longer in the Bremen collection (C. Hinkelmann *in litt*).]

Lesser Flamingo	*Phoeniconaias minor*	**VA;–;–**
	(Geoffroy 1798)	

Port: Flamingo-menor (C&C).

Afrotropical and S Asia; W Africa south to Cameroon, southwest Africa from Angola southwards (Brown *et al* 1982). One record Príncipe.

An immature male was obtained in June 1901 near Santo António by Fea, who was told that it was not rare and occurred in groups, 'fishing' on the rivers (Salvadori 1903a).

ANATIDAE

Knob-billed Duck	*Sarkidiornis melanotos*	**–;VA;–**
	(Pennant 1769)	

Port: Pato-de-carúncula, Pato-de-crista (F, C&C).

Pantropical, including Afrotropical resident (*S. m. melanotos*). Vagrant São Tomé.

One adult male collected near Rio Lembá, Dec 1954 (Frade 1958, Frade & Vieira dos Santos 1977).

ACCIPITRIDAE

Black (Yellow-billed) Kite	*Milvus migrans*	**RB;RB;V**
	(Boddaert 1783)	

Falcão (N 1972b, R, C&C). Port: Milhafre-preto, Rabo-de-bacalhau (N 1994, C&C).

Africa and Eurasia, migrant and resident. Resident São Tomé and Príncipe, vagrant Annobón. All birds, either seen clearly enough for identification by Jones & Tye (1988) and Eccles (1988), or else collected on the islands, were of the African race *parasitus*; Palearctic migrant *M. m. migrans* as yet unrecorded from the islands.

Príncipe Not seen by Dohrn (1866), nor apparently by Keulemans (1866), although the latter reported that it visited the S coasts at that time, probably coming from São Tomé. Keulemans reported that, on Príncipe, the people believed that the Grey Parrots attacked kites, sometimes killing them, or as Dohrn put it, 'there is a deadly hatred between the Grey Parrots (*Psittacus erythacus* [*sic*]) of Príncipe and the kites of San Thomé, and that, if ever a *Milvus* visits the neighbouring island, hundreds of Parrots fall upon him and kill him, and that the Kites take revenge if perchance a Parrot should venture a trip to their kingdom. There must be some family reason for this strange degree of enmity, for they seem to live in tolerable peace together on the coast [mainland].' Keulemans's record 'semble être passée inaperçue des auteurs ultérieurs' according to Naurois (1983b), presumably including himself (Naurois 1983a) among the 'auteurs ultérieurs'! Not seen again on Príncipe until 1954 (Frade 1958), and nests attributed to this species were found in 1970 by Naurois (1983b). By this time, the kites and parrots had apparently settled their differences (Naurois 1983b); perhaps Dohrn's (1866) and Keulemans' (1866) observations were based on a single incident. In 1987, frequent in habitats similar to those occupied on São Tomé (Jones & Tye 1988). Perhaps formerly an irregular visitor (Naurois 1983b) but now apparently an established resident, most common in the north and centre (Christy & Clarke 1998).

São Tomé Found by most of the early collectors, and apparently abundant about 1500 ('Beaucoup de faucons' of Piriz: Monod *et al* 1951) and in the mid-19th century (eg Hartlaub 1850, Dohrn 1866, Bocage 1891). Still abundant, along coast, in northern savannas, city suburbs and open areas throughout; common to abundant over forests and plantations in SE (Jones & Tye 1988). At least formerly, common in plantations at higher altitudes (Snow 1950), and occasionally up to 1000 m (Bocage 1904).

Annobón The only record is a bird collected by Newton between Nov 1892 and Jan 1893 (Bocage 1893b).

Habits Attracted to grass fires, where it captures fleeing lizards and insects (Jones & Tye 1988). Feeds at sea, following fishing canoes (J&T, Sargeant 1994). At least previously (Günther & Feiler 1985), hunted along roads for animals killed by traffic, although this could scarcely have been a profitable foraging tactic in 1987 (J&T). Also eats lizards, large crustaceans, dead fish, oil-palm fruits, and reportedly takes fruit bats *Eidolon helvum* (Naurois 1983b, Günther & Feiler 1985, J&T).

Breeding The nest and eggs were described by Naurois (1983b). The breeding season appears to be Jun–Nov (Amadon 1953, Naurois 1983b). One pair in flight-display, Apr (Nadler 1993).

FALCONIDAE

[Common Kestrel   *Falco tinnunculus*   ?;–;–
                             Linnaeus 1758

Afrotropical resident (*F. t. rufescens*) and Palearctic migrant (*F. t. tinnunculus*). One unconfirmed report from Príncipe.

Naurois (1975a) mentions in passing that a Common Kestrel was captured 'en migration' in 1974. This is the only record for the islands, and it was not mentioned again by Naurois (1983b). We therefore consider it unconfirmed.]

Western Red-footed Falcon   *Falco vespertinus*   –;VPM;–
                                       Linnaeus 1766

Falcão (F&V). Port: Francelho-de-pés-vermelhos (C&C).

Palearctic migrant to southern Africa; recorded on passgae from Canaries and W. Africa, Cameroon, Angola. Vagrant São Tomé.

Two individuals seen near the airport (NE), 26 Nov 1954 (Frade 1958, 1959); one of them (an immature male) was collected.

Peregrine Falcon   *Falco peregrinus*   –;VPM;–
                                Tunstall 1771

Port: Falcão-real, Falcão-peregrino (C&C).

Resident and Palearctic migrant to sub-Saharan Africa (Brown *et al* 1982). One record São Tomé.
    One adult of race *calidus* (N Eurasian migrant to E Africa) observed by P. Christy above Bom Sucesso, Dec 1994 (Christy & Clarke 1998, P. Christy *in litt*).

PHASIANIDAE

Feral Chicken   *Gallus gallus*   –;–;RB
                                (Linnaeus 1758)

Gañia, Gallina domestica (Bar, Bas).

Southeast Asia. Resident Annobón, having escaped from domestication.
    Apparently common in the forests in the 1950s (Basilio 1957). Reportedly smaller than European domestic chickens, but larger and more colourful than those in domesticity on the island (Barrena 1911, Basilio 1957). Domestic chickens were introduced by the Portuguese in the first years of settlement of the island (1503 ff.), as recorded by Valentim Fernandes (Monod *et al* 1951).

Helmeted Guineafowl   *Numida meleagris*   ?;RB;RB
                                (Linnaeus 1758)

STP: Galinha-do-mato. A: Gañia Guinee (Bar), Gallina de Guinea (Sp: Bar). Port: Pintada, Galinha-da-Guiné, Capota (N 1981, C&C).

Afrotropical; resident on São Tomé and Annobón, probably *N. m. galeata* and probably originally introduced to both; unconfirmed old reports from Príncipe.

Príncipe Perhaps once (or still?) occurred, as some local people there described a 'Galinha brava' living in the mountains, which seems to have been a guineafowl (Keulemans 1866).

São Tomé First recorded by the first Portuguese viceroy of India, Francisco d'Almeyda in (probably) 1505 (see Sousa 1888), and at about the same time by Valentim Fernandes, based on a report by Gonçalvo Piriz who, despite including them among domestic birds, said that they were very wild (Monod *et al* 1951). Collected by Weiss in 1847 (Hartlaub 1850). Newton considered it common in the north in the 1890s (Naurois 1981). Correia (1928–29b) found large flocks in tall razor-grass near Roça Boa Entrada (N) but collected only one male there and one elsewhere (Amadon 1953). Naurois (1981) found it uncommon in the savannas and in dense secondary growth near the forest edge, up to 200 m. Günther & Feiler (1985) and Nadler (1993) heard it in the N although Jones & Tye (1988) failed to find it; probably still exists at low density. Hunted for food, which may have contributed to a decline (Naurois 1981). An albino specimen exists in Coimbra Museum (Themido 1938) but could have been a captive bird.

Annobón Present in grassy areas in the early 19th century (Allen & Thomson 1848) and very common in 1892 and the 1950s in groups of 4 to 20, in open, cultivated areas throughout (Bocage 1903, Schultze 1913, Bannerman 1915b, Basilio 1957, Fry 1961). Still present in 1989 (Harrison 1990).

Habits Food includes seeds and grasshoppers (Basilio 1957).

Breeding Unrecorded. Many specimens collected for Basilio (1957) in Sep and Oct had small gonads; in Nov some had enlarged ones and Basilio was told that laying occurred in Dec and young were numerous in Jan. The nest was described by Basilio (1957), who also reported many nests destroyed by burning of grasslands.

Harlequin Quail	*Coturnix delegorguei histrionica*	–;RB*;–
	Hartlaub 1849	

Codorniz (Keulemans *in* Sharpe 1869, R). Port: Codorniz de São Tomé (N 1994); Codorniz-arlequim (C&C).

Subspecies endemic to São Tomé; others on African mainland (vagrant Bioko), Arabia and Madagascar.
    Considered common but secretive by Keulemans, in grasslands and swamps near the capital in the 1860s (Sharpe 1869). Naurois & Castro Antunes (1973) show it in the northern savannas and the SE. More recently, frequent in Ribeira Peixe oil-palm plantation (SE) in groups of up to 12 (Jones & Tye 1988); frequent on farmland in the N and E, whether fallow or bearing crops (Bocage 1891, 1904, Bannerman 1915a, Günther & Feiler 1985, Eccles 1988, Jones & Tye 1988); also found commonly in tall-grass savanna (Correia 1928–29b, Günther & Feiler 1985, Nadler 1993). Naurois (1972b, 1981) and Nadler (1993) mention its presence in coconut plantations but, since independence, the heavy development of undergrowth in many of these would probably have resulted in its removal. Once heard in the mountains (Keulemans *in* Sharpe 1869); once on a football pitch (Nadler 1993).

Habits Stomach contents include seeds, beetles and caterpillars (Sharpe 1869, Naurois 1981). The call and habits were described by Keulemans (*in* Sharpe 1869) and Nadler (1993). Alarm call when flushed is a rusty squeak *eek ik ik ik ik ik* (J&T). A sonogram was presented by Nadler (1993).

Breeding The egg was described by Bocage (1891, 1904) and Naurois (1994), and the nest by Naurois (1981). Considered probably breeding in early Feb by Alexander (Bannerman 1915a) and Apr by Newton (Naurois 1981) while specimens collected by Correia, May–Jul, were mostly in breeding condition (Amadon 1953). Naurois (1972b) stated that laying period was Dec–Jan, but gave no evidence. Two clutches: Mar, Oct (Naurois 1981, 1994).

Morphometrics of a trapped bird given by Atkinson *et al* (1994).

Red-necked Spurfowl	*Francolinus afer*	–;RB;–
	(Müller 1776)	

Perdiz (C&C). Port: Perdiz-de-gola-vermelha (C&C).

Afrotropical. Resident São Tomé.
    Tentatively recorded for the first time (sightings) between 700 and 900 m above Nova Moca by S. Reinius (*in litt*), who initially attributed the birds to *F. squamatus* (Reinius 1985). Local people also informed Reinius that a similar (or the same) bird occurred in the lowlands. In 1987 an individual was seen clearly on the path between Lembá and Bindá, W coast (Jones & Tye 1988). Rustlings in the undergrowth nearby suggested that other individuals were present. One heard calling in the same place the following year (J&B). One at Boca do Inferno (extreme S), Aug 1990 (Atkinson *et*

al 1994). Two seen *c* 1.5 km NW of the airport in Oct 1990 (S. Reinius *in litt*); these birds fitted the description of the Angolan race *F. a. afer*. More recently, shown to be widespread in savannas of NW, cocoa plantations in the W and E, and up to at least 1000 m near Bom Sucesso (Christy & Clarke 1998). Apparently now a widespread resident, its origin and period of residence on the island are unknown.

RALLIDAE

African Crake   *Crex egregia*   **VA;VA;–**
                             (Peters 1854)

Tôghe (Newton in N 1972b); Codornizão (Correia 1928–29b, F&V). Port: Codornizão-africano (C&C).

Afrotropical including Bioko. Probable vagrant, Príncipe and São Tomé.

Príncipe A very thin adult female collected in Dec 1970, and a similar bird reported there the previous winter (Naurois 1987a). Recorded, without details, near the dependência de Bela Vista in Jan (Christy & Clarke 1998).

São Tomé Two adult females were collected at Rio do Ouro and Campo de Santo Amaro (N and NE), by Newton (Bocage 1889c, 1891, 1904); Newton reported that it had been more common in earlier times and had gained a local name (Naurois 1987a). Correia collected another at Santo Amaro in a marsh of grass and sugar cane (Correia 1928–29b, Amadon 1953), and Frade a male at Mogadinho, Sep 1954 (Frade & Vieira dos Santos 1977). Some of the specimens were reportedly very thin, evidently stragglers from the continent (Naurois 1983a).

African Water Rail   *Rallus caerulescens*   **–;VA;–**
                             Gmelin 1789

Port: Frango-d'água (C&C).

Afrotropical. Vagrant São Tomé.

A single specimen was sent by Gomes Roberto to the King, D. Pedro V, in 1861 with a collection of birds from the island (Bocage 1867, 1904, Sousa 1888). There is no doubt about the origin of the specimen (Sousa 1888, Bocage 1904) but Bocage (1904) regarded its existence as a resident unproven, because this was the only record.

Allen's Gallinule   *Porphyrio alleni*   **VA;VA(B?);VA**
                             (Thomson 1842)

Port: Sultana-preta, Galinha-sultana-pequena (F&V, C&C).

Afrotropical including Bioko; undertakes intra-African migrations. Vagrant Príncipe, São Tomé and Annobón.

Príncipe An adult sent to Frade in Lisbon in 1956 is now at IICT (Naurois 1987a).

São Tomé First recorded when a juvenile female was collected by Alexander, Feb 1909 (Bannerman 1915a). A male was obtained by Frade at Mogadinho, Sep 1954 (Frade & Vieira dos Santos 1977), and a fully-grown immature on the Lagoa de Malanza by Naurois (1972b, 1987a). One was seen in 1987 in a lagoon near Praia Micoló (Eccles 1988).

Annobón Known from a single exhausted individual captured at the northern tip of the island, Aug 1959 (Fry 1961). This is probably the specimen at MNCN, which is an immature (J. Pérez del Val *in litt*).

Common Moorhen	*Gallinula chloropus*	**RB;RB;E**
	(Linnaeus 1758)	

STP: Galo d'aua (B 1889c); Galinha-d'água (N 1994). A: Quelvo (Bar, Bas); Polla de agua africana (Sp: Bas). Port: Galinha-de-água, Galinha-de-água-africana (F&V, N 1994).

Near-cosmopolitan. *G. c. meridionalis* in sub-Saharan Africa. Resident São Tomé and Príncipe, extinct Annobón.

Príncipe Frequent along lower reaches of rivers and in marshland (Naurois 1987a, Jones & Tye 1988). First discovered by Frade in 1956 (Frade & Vieira dos Santos 1977; Naurois 1972b, 1983a, 1987a gave discovery date as 1954). Naurois (1972b) first found it breeding.

São Tomé First collected by Weiss (Hartlaub 1855) and recorded by most subsequent ornithologists (Naurois 1987a). Now common, usually in pairs, along lower reaches of rivers and in marshes and lakes, and occasionally heard in undergrowth of open areas far from water in SE, sometimes at night (Naurois 1983a, 1987a, Jones & Tye 1988).

Annobón This species may have been the 'wild duck' recorded by Valentim Fernandes as common on the crater lake, in the early years of the 16th century (Monod *et al* 1951). Lived only on the lake, where it was rare in 1892 (Bocage 1903); collected in a 'reedbed' of *Polygonum* by Alexander in 1909 (Bannerman 1915b, Fry 1961), and last recorded in December 1910 (Lowe 1932, BMNH registers). Now apparently extinct; said to have died out only a few years prior to 1955, perhaps exterminated by a combination of nest-robbing by boys and Reef Herons *Egretta gularis*, or by clearing the lake-side vegetation for mosquito control, so depriving it of nesting habitat (Basilio 1957).

Habits Stomach contents included small seeds, beetles and fruit pulp (Naurois 1987a).

Breeding On Annobón said to build its nest on the water, fixed to vegetation (Barrena 1911). Nest and eggs described by Naurois (1987a, 1994), who gave breeding season as Jul to early Jan; eggs laid Jul, Sep, Nov–Dec (Naurois 1987a).

Lesser Moorhen	*Gallinula angulata*	**VA;RB?;–**
	Sundevall 1850	

Galla d'aua (S 1903b); Frango-de-água (F&V). Port: Galinha-d'água-pequena (N 1994).

Afrotropical. Probable vagrant Príncipe, perhaps resident São Tomé.

Príncipe A male collected at Santo António, Jun 1956 (Frade & Vieira dos Santos 1977) (in 1954 according to Naurois 1972b but not mentioned by Naurois 1987a, where he instead reported one from São Tomé, probably in error).

São Tomé A specimen was sent to the King, D. Pedro V, in 1861 by Gomes Roberto, in a collection of birds from the island (Sousa 1888, see also Bocage 1904). Günther & Feiler (1985) reported finding a 19th-century specimen (no collector listed) from São Tomé in the Museum Heineanum. This is presumably the specimen listed for that museum by Heine & Reichenow (1882–90) and Salvadori (1903b). Naurois (1983a) reported 'sérieux indices de reproduction' (elaborated as a nest, smaller and less thick than that of Common Moorhen *G. chloropus*, by Naurois 1994, with other details given by Naurois 1987a) in 1971–72 on São Tomé; however, the evidence detailed by Naurois in these publications is inconclusive, and eggs have never been found (Naurois 1994). Naurois (1983a) also suggested that it had colonised the island since 1929 but the 19th-century record, together with the fact that the bird is retiring and easily mistaken for Common Moorhen, allows that it may have been present before then.

GLAREOLIDAE

Black-winged Pratincole   *Glareola nordmanni*   **VPM;–;VPM**
                                Fischer 1842

STP: Perdiz-do-mar (F&V). A: Canastera de Nordmann (Sp: Bas). Port: Pratíncola-de-asa-preta (C&C).

Eurasian migrant wintering in southern Africa. Known from Príncipe and Annobón, only 4 records.

Príncipe Three specimens were collected: a male in Sep 1865 (Dohrn 1866, Keulemans 1866), a female by Correia in Sep 1928 (Amadon 1953, L.R. Macaulay *in litt*) and another female by Frade in Nov 1954 (Frade & Vieira dos Santos 1977).

Annobón A pair was seen in Sep–Oct 1955; the male was shot (Basilio 1957) and the other disappeared in Nov (Basilio 1957). They were seen mainly when hawking insects above the town at dusk.

Habits Stomach contents consisted of grasshoppers (the Príncipe male: Keulemans 1866), small beetles and green Hemiptera (the Annobón bird: Basilio 1957).

CHARADRIIDAE

[Little Ringed Plover   *Charadrius dubius*   –;?;–
                              Scopoli 1786

Palearctic migrant to tropical Africa (*C. d. curonicus*) and S Asia.

São Tomé One reported at Praia Melão, 26 and 28 Dec 1984 (Reinius 1985 and *in litt*).]

[Ringed Plover   *Charadrius hiaticula*   –;?;–
                       (Linnaeus 1758)

Port: Borrelho-de-coleira (C&C).

Breeds Holarctic, winters to Africa, including Bioko.

São Tomé Two immatures seen, Jan 1995 at Praia dos Tamarindos (Christy & Clarke 1998) or Praia das Conchas (P. Christy *in litt*). Without further details, this record must be regarded as provisional.]

White-fronted Sand Plover *Charadrius marginatus* –;VA;–
Vieillot 1818

Borrelho (F). Port: Borrelho-de-fronte-branca (C&C).

Afrotropical, including Gulf of Guinea coast.

São Tomé A small flock, of which one male was captured, was seen several times at the airport (NE), Nov 1954 (Frade 1958, Frade & Vieira dos Santos 1977). Subspecies not recorded.

'Lesser' Golden Plover *Pluvialis sp.* –;V;–
Port: Tarambola-dominicana (F&V, C&C).

São Tomé Two individuals seen, of which an immature female was collected, at the airport (NE), 24 Nov 1954 (Frade 1958, Frade & Vieira dos Santos 1977). The identity of these birds is uncertain, given that both Pacific Golden Plover *P. fulva* and American Golden Plover *P. dominica* have been recorded in Gabon (Christy 1990, P. Alexander-Marrack, pers comm) and Ivory Coast (Fishpool & Demey 1991). Note that the birds recorded by Frade (Frade 1958, 1959; see also Günther & Feiler 1985) were initially referred to Eurasian Golden Plover *P. apricaria* (Frade & Vieira dos Santos 1977).

[Eurasian Golden Plover *Pluvialis apricaria* –;?;–
(Linnaeus 1758)

São Tomé Reportedly seen at the airport in Sep–Oct 1997 by P. Kaestner (*per* P. Atkinson *in litt*).]

Grey Plover *Pluvialis squatarola* –;VPM;–
(Linnaeus 1758)

Tarambola-cinzenta (Port: F&V).

Arctic migrant to S hemisphere; wintering areas include entire African coast.

São Tomé One female collected at Diogo Nunes (NE), 8 Mar 1970 (Frade & Vieira dos Santos 1977). One seen on the shore in the city and *c* 20 around an inland marsh (NE), Feb 1989 (Harrison & Steele 1989).

SCOLOPACIDAE

Sanderling *Calidris alba* –;–;PM
(Pallas 1764)

Chunchu (Bar, Bas); Correlimos tridáctilo (Sp: Bas).

Arctic migrant to S hemisphere; wintering areas include entire African coast. Uncommon visitor Annobón.

Recorded by Barrena (1911: probably this species), and in 1955 by Basilio (1957) who noted groups of 2, 3 and up to 7 on beaches in Oct, mainly on the W coast. One collected had sand, shell fragments and soft parts of molluscs in the stomach. Local people reported its presence throughout the year, and existence of a local name suggests it is a regular migrant (Basilio 1957). This is one of the few species of wader likely to migrate in large numbers through the Gulf of Guinea to the south-west African coast (Summers & Waltner 1979, Tye 1987).

[Stint sp   *Calidris* sp   **?;?;–**

Príncipe Keulemans (1866) noted 3 species of sandpiper, 2 of which were probably the Greenshank *Tringa nebularia* and Curlew Sandpiper *Calidris ferruginea* (*qv*). His third species was the smallest and so probably refers to a species of stint.

São Tomé A stint was seen in 1988 (J&B), but was not identified. **]**

Pectoral Sandpiper   *Calidris melanotos*   **V;–;–**
                                    (Vieillot 1819)

Port: Maçarico-de-costas-pretas (F&V, C&C).

Siberian and Nearctic migrant to Australasia and S America. Vagrant Príncipe, origin unknown.

One immature male collected in Baía de Santo António, 14 Nov 1954 (Frade 1958; the specimen was termed 'adult' by Frade 1959).

Curlew Sandpiper   *Calidris ferruginea*   **VPM;VPM;–**
                                    (Pontoppidan 1763)

Port: Pilrito-de-rabadilha-branca (C&C).

Palearctic migrant to Africa, S Asia and Australia. One record São Tomé; rare Príncipe.

Príncipe Recorded by Dohrn (1866: '*Tringa subarquata*'), apparently throughout the year (Bannerman 1914), as in other parts of Africa (eg Summers *et al* 1977, Urban *et al* 1986, Tye & Tye 1987). One collected in Jun or Jul was in 'winter' plumage (Dohrn 1866), and was presumably an immature bird. Keulemans' (1866) sandpiper with the moderately long, down-curved beak may have been this species but Correia's (1929) 'snipe' with short curved beak was a Black-winged Pratincole *Glareola nordmanni* (Amadon 1953, L.R. Macaulay *in litt*). More recently, recorded in the bay at Santo António, Jan 1995 (Christy & Clarke 1998).

São Tomé Two at Morro Peixe, 27–28 Aug 1990 (Atkinson *et al* 1994). One at Praia das Conchas 23 Nov 2004 (C. Hjort *in litt*).

Bar-tailed Godwit   *Limosa lapponica*   **VPM;VPM;VPM**
                                    (Linnaeus 1758)

Port: Fuselo, Parda (C&C).

Palearctic migrant to entire coastline of Africa (*L. l. lapponica*), SE Asia and Australia. Vagrant Príncipe, São Tomé and Annobón.

Príncipe One in the harbour at Santo António, Jan 1995 (Christy & Clarke 1998).

São Tomé Two on a puddle on a wet grassy path at São Miguel, Nov 1984 (M. Goulding *in litt*).

Annobón. One record in August 2000 (Pérez del Val 2001)

Whimbrel   *Numenius phaeopus*   **PM;PM;PM**
                                (Linnaeus 1758)

STP: Côco-piloto (B 1889c, 1891). A: Hoho (B 1893b); Jojo (pron. Hoho) pilot (Bar); Jojó pilot (Bas); Zarapito menor (Sp: Bas). Port: Meio-maçarico, Maçarico-galego (F&V, C&C).

Arctic migrant to S hemisphere, including entire African coast (*N. p. phaeopus*) and all 3 islands.

São Tomé and Príncipe Frequent to common from early Aug (Correia 1928–29b, Jones & Tye 1988) to at least Apr (Nadler 1993). Oversummers in small numbers (Keulemans 1866, Günther & Feiler 1985). Inhabits rocky and muddy coasts and occasionally occurs up rivers (Jones & Tye 1988).

Annobón Newton collected a female on Tortuga islet between Nov 1892 and Jan 1893 (Bocage 1893b). Familiar to Barrena (1911). Basilio (1957) recorded *c* 7 in autumn 1955, singly or in groups of 2–3 near Palea and on the beach at Áquequel. Basilio was told by local people that they were present all year, but never bred.

Eurasian Curlew   *Numenius arquata*   **VPM;VPM;–**
                                (Linnaeus 1758)

Port: Maçarico-real (C&C).

Palearctic migrant to tropical Africa and SE Asia, including entire African coast; *N. a. arquata* in W Africa, although island birds referred to *orientalis* by Christy & Clarke (1998). Vagrant São Tomé and Príncipe.

Príncipe Recorded by Dohrn (1866), although Keulemans (1866) saw only Whimbrel *N. phaeopus*. Following Salvadori (1903a) we consider Dohrn's record of Eurasian Curlew doubtful. However, one was clearly seen, with a Whimbrel, on the Rio Banzul, 29 July 1988 (J&B).

São Tomé Correia collected a female, 14 Aug 1928 (L.R. Macaulay *in litt*); also reported Jul, without details (Christy & Clarke 1998).

Greenshank   *Tringa nebularia*   **PM;PM;–**
                                (Gunnerus 1767)

Port: Perna-verde, Maçarico-cinzento (C&C).

Palearctic migrant to tropical Africa, S Asia and Australasia. Uncommon São Tomé and Príncipe, on estuaries and lagoons.

Príncipe Seen by Dohrn (1866); this may have been the largest of the 'three kinds of sandpiper' seen by Keulemans (1866). Also seen by Reinius (1985 and *in litt*), and one seen by PJJ on the Rio Banzul (SW), 29 Jul 1988. Reportedly more common Dec–Jan (Christy & Clarke 1998).

São Tomé Seen by Reinius on the N coast (1985 and *in litt*); one near city, 22 Aug and one at Morro Peixe, 27–28 Aug (Atkinson *et al* 1994); one 11 Aug (Christy & Clarke 1998). One at mouth of Rio Lembá 24 Nov 2004 (C. Hjort *in litt*). Reportedly more common Dec–Jan (Christy & Clarke 1998).

[Green Sandpiper   *Tringa ochropus*   –;?;–
                                        Linnaeus 1758

Port: Pássaro-bique-bique (C&C).

Palearctic migrant to Africa and S Asia.

São Tomé Recorded at Praia dos Tamarindos on the N coast, Dec 1983 (Reinius 1985 and *in litt*). This record requires confirmation.]

Wood Sandpiper   *Tringa glareola*   –;PM;–
                                        Linnaeus 1758

Port: Maçarico-silvestre (C&C).

Palearctic migrant to Africa, S Asia and Australasia. Uncommon São Tomé.
    Two specimens obtained by Newton, at Diogo Nunes and Rio do Ouro (N) (Bocage 1889c, 1904); seen on the N coast by Reinius (1985 and *in litt*); 7 seen on coastal lagoons in the N (Eccles 1988); 3 seen at the lagoon at Praia das Conchas, Jan 1995 (P. Christy *in litt*).

Common Sandpiper   *Actitis hypoleucos*   PM;PM;–
                                        (Linnaeus 1758)

Maçarico (K). Port: Maçarico-das-rochas (C&C).

Palearctic migrant to Africa, S Asia and Australasia. Regular São Tomé and Príncipe.

São Tomé and Príncipe Frequent to common, Jul–Apr, on lower reaches of rivers and along the shore (Keulemans 1866, Bocage 1891, 1904, Snow 1950, Eccles 1988, Jones & Tye 1988, Nadler 1993, L.R. Macaulay *in litt*). Also recorded on Príncipe in swamps (Dohrn 1866) and on Ilhéu das Rolas (Bocage 1891, 1904).

Ruddy Turnstone   *Arenaria interpres*   VPM;VPM;–
                                        (Linnaeus 1758)

Port: Rola-do-mar (C&C).

Arctic migrant to S hemisphere; wintering areas include entire African coast (*A. i. interpres*). Uncommon Príncipe and São Tomé. The map in Urban *et al* (1986) incorrectly shows Ruddy Turnstone on Annobón: there is no evidence for its occurrence there (nor was there for Príncipe at the time).

Príncipe Reported on Pedras Tinhosas, without details (Christy & Clarke 1998).

São Tomé Two specimens collected by Newton at Fernão Dias (N) about 1887–88 (Bocage 1889a) and one on Ilhéu das Rolas in 1890 (Bocage 1891). Four individuals seen in non-breeding plumage on the shore north of Neves, Jun 1983 (Günther & Feiler 1985).

STERCORARIIDAE

Pomarine Skua   *Stercorarius pomarinus*   –;VPM;–
(Temminck 1815)

Port: Moleiro-pomarino, Mandrião (C&C).

Holarctic migrant to southern oceans, including W African coast to W Nigeria, S African coast from Namibia southward (Urban *et al* 1986) and seas west of the islands (5°W to 2°E) (Lambert 1988); reported common off Luanda (Angola) during the northern winter (M. Goulding *in litt*). Vagrant São Tomé.

One sighted off the N end, Nov 1984 (M. Goulding *in litt*): the bird flew over the observer at *c* 15 m altitude.

[Arctic Skua   *Stercorarius parasiticus*   ?;?;?
(Linnaeus 1758)

Port: Moleiro-parasítico (C&C).

Gañia sojofs (Bar, perhaps misprint for sojofa); Págalo parásito (Sp: Bas).

Holarctic migrant to southern oceans; recorded on passage off W Africa (Urban *et al* 1986) and in seas west of the islands (5°W to 2°E) (Lambert 1988). Uncertainly described for Annobón.

Basilio (1957) and Fry (1961) believed a bird to be this species, which was described by inhabitants as 'present offshore all year, never coming to land, of dark colours, and constantly attacking terns to steal fish from them'. Barrena's (1911) description amounts to no more than 'seabird'. There is no firm evidence of its identity. Seven reported by I. Sinclair at sea between São Tomé and Príncipe, Mar 1992 (P. Christy *in litt*). These records require confirmation.]

[Long-tailed Skua   *Stercorarius longicaudus*   ?;?;?
Vieillot 1819

Breeds Holarctic, winters mainly in southern oceans; one record Gulf of Guinea, west of São Tomé (0°17'N 3°24'E) (Urban *et al* 1986).

Barrena's (1911) 'Quinsá' might possibly refer to this species. Two adults in breeding plumage seen by I. Sinclair between São Tomé and Príncipe, 30 Mar 1992 (Christy & Clarke 1998, P. Christy *in litt*). These records require confirmation.]

LARIDAE

[Lesser Black-backed Gull   *Larus fuscus*   –;–;–
Linnaeus 1758

Breeds Eurasia, winters south to Africa, including W African coast.

According to P. Christy (*in litt*), reported for São Tomé and/or Príncipe in 1949 by W.R.P. Bourne. However, Bourne himself (*in litt*) referred these reports to unidentified gulls listed by Cadée (1981), adding only that the 'most likely' species was *L. fuscus*; in addition, the possibility of occurrence of Kelp Gull *L. dominicanus* suggests that this record still requires confirmation.]

[Sabine's Gull   *Larus sabini*   ?;?;?
           Sabine 1819

Port: Gaviota-de-Sabine (C&C).

Holarctic migrant to southern oceans; common at migration time in seas west of the islands (5°W to 2°E) (Lambert 1988); wintering areas include southwest African coast. Reported at 0°2'S 5°12'E and 0°58'S 7°40'E, in the seas around São Tomé and Annóbon, Nov 1979 (Cadée 1981). Also reported at sea near São Tomé and/or Príncipe in 1949 by the Oxford Expedition (W.R.P. Bourne *per* P. Christy *in litt*). Although likely to occur, these records require confirmation.]

[Royal Tern   *Sterna maxima*   ?;?;–
          Boddaert 1783

Port: Andorinha-do-mar-real, Gaivina-real (C&C).

Atlantic and Pacific; *S. m. albididorsalis* breeds W Africa.
  Four reported by I. Sinclair between São Tomé and Príncipe, Mar 1992 (Christy & Clarke 1998, P. Christy *in litt*). This record requires confirmation.]

Sandwich Tern   *Sterna sandvicensis*   –;PM;–
          Latham 1787

Port: Garajau, Andorinha-do-mar de Sandwich (C&C).

Holarctic migrant to the tropics; wintering areas include entire African coast (*S. s. sandvicensis*). Regular visitor to São Tomé.
  Often seen in small groups of 3–5 birds in the northern winter in the mid-1980s (S. Reinius *in litt*); 3, in breeding dress, seen 17 Apr 1991 (Nadler 1993). Two ringed as nestlings, one in England, Jun 1975, the other in Holland, Jul 1975, were recovered on São Tomé on 11 Nov that year by the same finder, presumably a tern hunter (Mead & Clark 1987, C.J. Mead *in litt*, R. Wassenaar *in litt*). A third nestling, ringed in Holland in Jun or Jul 1983, was recovered in Nov 1983 by Reinius (1985 and *in litt*).

Common Tern   *Sterna hirundo*   ?;?;–
          Linnaeus 1758

Holarctic, wintering to southern oceans; frequent at migration time west of the islands (5°W to 2°E) (Lambert 1988). May be a regular visitor to the islands, but confirmation of the identifications is required (see next species).
  Birds of this or the next species were often seen in small groups during the northern winter in the mid-1980s (Reinius 1985 and *in litt*). Common/Arctic Terns were also seen between São Tomé and Sete Pedras, and between São Tomé and Príncipe, Mar; off the north coast of São Tomé, Dec; and around Pedras Tinhosas, Jan and Aug (Christy & Clarke 1998). The Dec–Jan observations at least were probably Common Tern (Christy & Clarke 1998). Resting birds identified at close range at Lagoa Azul, N. Coast of São Tomé, 24 Nov 2005 (C. Hjort *in litt*).

[Arctic Tern   *Sterna paradisaea*   ?;?;–
                             Pontoppidan 1763

Holarctic, wintering to southern oceans; frequent at migration time west of the islands (5°W to 2°E) (Lambert 1988). May be a regular visitor to the islands, but confirmation of the identifications is required (see previous species).

Birds of this or the previous species were often seen in small groups during the northern winter in the mid-1980s (Reinius 1985 and *in litt*). Common/Arctic Terns were also seen between São Tomé and Sete Pedras, and between São Tomé and Príncipe, Mar; and around Pedras Tinhosas, Aug (Christy & Clarke 1998). Several Arctic Terns reportedly seen by P. Kaestner (*in litt* to P. Atkinson) Sep–Oct 1997.]

Bridled Tern   *Sterna anaethetus*   (B?)S;(B?)S;BS
                             Scopoli 1786

STP: Rabo-de-tzoura (K); Coco-sandjia (B 1891, 1904); Caié (Monteiro *et al* 1997). A: Guiyina (Bar); Gayin, Charrán de Guinea (Sp) (Bas).

Tropical and subtropical oceans. Breeds Annobón and probably also on the other islands.

Príncipe Recorded off the W coast by Dohrn (1866: '*S. melanoptera*') and Keulemans (1866: '*S. panayense*'), the latter reporting it uncommon (see Bocage 1903). One at Praia Grande, Aug 1991 (Sargeant 1994). Mostly recorded Jan and Aug (Christy & Clarke 1998). Gascoigne (1995) listed it as breeding on Pedras Tinhosas but presented no evidence; considered irregular there by Christy & Clarke (1998).

São Tomé Collected on the islets of Sete Pedras, where probably breeds (Bocage 1891, 1904); *c* 30 birds there (Naurois 1973a), several with large gonads collected (*per* P.L. Britton *in litt*); only 2 there, Aug 1990 (Atkinson *et al* 1994); some 60 there according to Christy & Clarke (1998: no details presented); up to 10 roosting there, Jan, Mar, Jul 1997 (Monteiro *et al* 1997). Seen fishing inshore off Ribeira Peixe (SE) within sight of Sete Pedras (Jones & Tye 1988), off Santo António (SW) (Atkinson *et al* 1994) and off Ilhéu das Cabras and E coast of main island (Monteiro *et al* 1997). Possibly breeds elsewhere on islets off São Tomé and Príncipe.

Annobón A colony of about 200 pairs (in 1959) bred or breeds on the islet of Tortuga, 400 m offshore (Basilio 1957, Fry 1961) but not seen there in 1964 (May–Jun) or 1965 (May) by Robins (1966). A male had moderately enlarged gonads, Sep (Basilio 1957).

Sooty Tern   *Sterna fuscata*   BS;(B?)S;(B?)S
                            Linnaeus 1766

Port: Gaivina-fosca, Andorinha-do-mar-escura (F&V, N 1994, C&C).

Tropical and subtropical oceans; *S. f. fuscata* in Africa. Breeds Príncipe and probably also on the other islands.

Príncipe Breeds on the islets of Pedras Tinhosas (Plate 32): several specimens, including juveniles, collected Mar, but breeding numbers (5000–200,000) given by Britton (*in* Urban *et al* 1986) have not been critically determined (R. de Naurois *per* P.L. Britton *in litt*): 600,000–1,200,000 pairs according to Naurois (1973a); several hundred immatures present Aug (Christy 1995b). A nestling was collected on Pedra da Galé, 4 km NW of Príncipe, Nov (Frade & Vieira dos Santos 1977), where regularly seen by Naurois (1973a) and up to 25 seen feeding, Aug (Sargeant 1994). However, Naurois

(1973a) suggested that the tiny, wave-washed Galé would rarely allow a nest to reach full term, and none seen there in Jul 1997 (Monteiro *et al* 1997).

São Tomé Collected by Ansorge, Jul (Bannerman 1930–51 vol. 2). Newton collected one on a boat 25 miles off the coast (Bocage 1889c). Reported at sea off the island, Sep 1949 (D.W. Snow *per* W.R.P. Bourne *in litt*). Recorded at sea 50 km NNE of São Tomé and between São Tomé and Sete Pedras; commoner nearer Príncipe than São Tomé (Fry 1961, Christy & Clarke 1998).

São Tomé & Annobón Reported at sea at *c* 0°2′S 5°12′E, 0°13′S 5°15′E and 0°58′S 7°40′E, Nov 1979 (Cadée 1981).

Annobón First recorded in May 1964 or 1965, when they were the most common seabird on Tortuga islet (100s), and 100–200 were also present on the southern islets (Adams, Santarém, Escobár, Fernando Póo) (Robins 1966). Their presence may be seasonal, accounting for their having been missed by Fry (1961).

Habits Feeds far out to sea; the feeding grounds of the enormous Tinhosas population remain to be discovered, the waters of the Gulf of Guinea being comparatively nutrient-poor (Naurois 1973a).

Breeding Nests and eggs described by Naurois (1973a) and Monteiro *et al* (1997); the length of the reproductive cycle (annual or less) is unknown. On Tinhosa Grande, mixture of adults with eggs, nestlings and fledglings at all stages, Aug 1995; many adults incubating eggs, few chicks, Jan 1996; few eggs and chicks but more fledglings Apr 1997; eggs and young chicks Jun–Jul; eggs and young chicks Sep–Oct 1997 (P. Kaestner *in litt* to P. Atkinson); mostly suggesting two laying peaks in Dec–Jan and May–Jun (Monteiro *et al* 1997, Christy & Clarke 1998).

[Black Tern	*Chlidonias niger*	?;?;–
	(Linnaeus 1758)	

Port: Gaivina-preta (C&C).

Holarctic migrant to Africa (*C. n. niger*) and S America; frequent at migration time west of the islands (5°W to 2°E) (Lambert 1988). It was listed, without details, for Bioko (Pérez del Val *et al* 1994).

The map in Urban *et al* (1986) shows this species occurring around all three islands, but there is no firm evidence that it does so, the only reports being of immature birds 31 Dec 1992 (P. Christy) and 30 Mar 1992 (200 individuals: I. Sinclair) between São Tomé and Príncipe, and 'thousands' at sea, Apr 1995 (I. Sinclair) (Christy & Clarke 1998, P. Christy *in litt*).

São Tomé: 30 reportedly seen by I. Sinclair between the city and Sete Pedras, 28 Mar 1992 (P. Christy *in litt*). Reportedly seen by P. Kaestner (*in litt* to P. Atkinson) in Sep–Oct 1997, but no details of the sighting are available.]

White-winged Black Tern	*Chlidonias leucopterus*	–;–;VPM
	(Temminck 1815)	

Palearctic migrant to Africa, S Asia and Australia. One record Annobón.

A juvenile observed on the crater lake, Aug 1959 (Fry 1961).

Plate 1 View south from the Pico de São Tomé over the undisturbed forested interior (*obó*) of São Tomé (Martim Melo)

Plate 2 Lowland forest in southern São Tomé (Peter Jones)

Plate 3 Cão Grande (663 m) near Ribeira Peixe, southern São Tomé (Peter Jones)

Plate 4 Carriote (840 m) from the Ribeira de São Tomé, southern Príncipe (Martin Dallimer)

Plate 5 João Dias Filho e Pai, Rio Banzúl, Príncipe (Peter Jones)

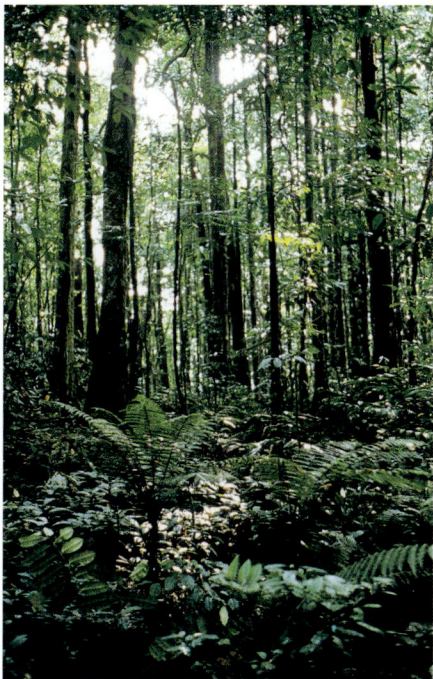

Plate 6 Lowland forest interior, Príncipe (Martin Dallimer)

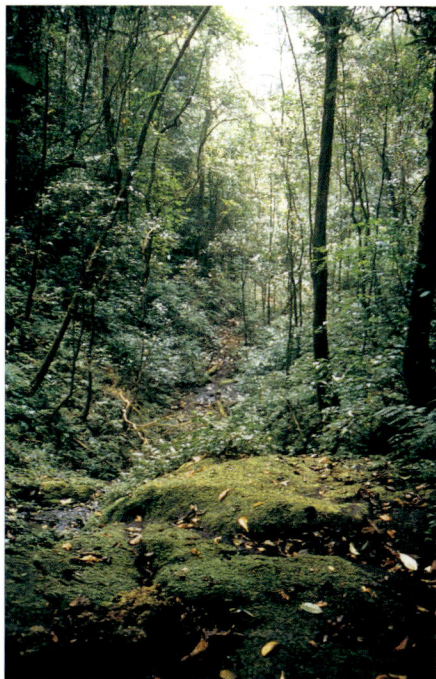

Plate 7 Montane forest interior, São Tomé (Martim Melo)

Plate 8 Mist-forest interior, São Tomé (Martim Melo)

Plate 9 Vegetation-filled surface of Lagoa Amélia and montane forest on the crater rim, São Tomé (Peter Jones)

Plate 10 Montane forest cleared for cultivation, Bom Sucesso, São Tomé (Peter Jones)

Plate 11   View over undisturbed montane forest towards Pico de Ana Chaves (1630 m), São Tomé (Martin Dallimer)

Plate 12   Closed canopy of shade forest (*floresta de sombra*) over cocoa plantation at Porto Real, Príncipe, showing flowering *Erythrina* (Peter Jones)

Plate 13   Interior of shade coffee plantation (*floresta de sombra*), São Tomé
(Martin Dallimer)

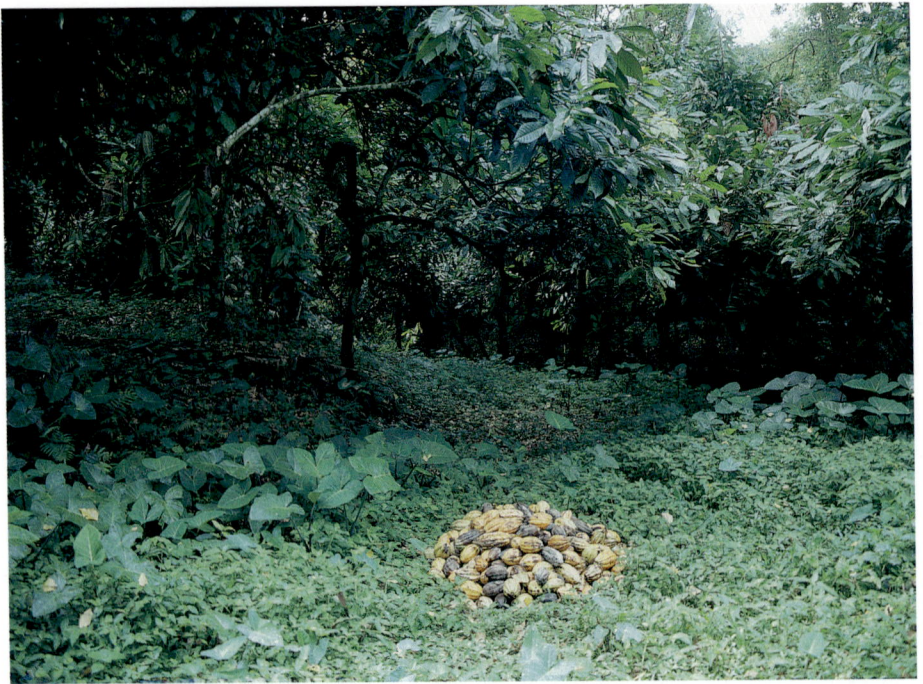

Plate 14   Interior of shade cocoa plantation (*floresta de sombra*), Príncipe
(Peter Jones)

Plate 15 View west on the northern coast of São Tomé at Lagoa Azul, showing savanna grassland with baobabs (Peter Jones)

Plate 16 View east on the northern coast of São Tomé at Morro Peixe, showing tidal lagoon and coconut plantations (Peter Jones)

Plate 17   Dry season savanna landscape in northern São Tomé (Peter Jones)

Plate 18   Newly burnt savanna grassland in northern São Tomé (Alan Tye)

Plate 19 Degraded mangrove habitat in tidal lagoon at Rio Malanza, southern São Tomé (Peter Jones)

Plate 20 Mudflats in tidal lagoon at Morro Peixe, northern São Tomé (Peter Jones)

Plate 21 Lower section of the Rio Io Grande, São Tomé's largest river (Peter Jones)

Plate 22 Wetland habitat on the Rio Lembá, western São Tomé (Alan Tye)

Plate 23   Ôque Pipi (325 m) and degraded farmbush, Príncipe (Peter Jones)

Plate 24   Abandoned estate buildings at Maria Correia, southwest Príncipe
(Martim Melo)

Plate 25   Beach at Praia Seca, Príncipe (Martim Melo)

Plate 26   *Pandanus candelabrum* on the estuary of the Ribeira de São Tomé, southern Príncipe (Martin Dallimer)

Plate 27 View of southeastern Príncipe towards the Pico do Príncipe (948 m) on the right, the 'twin' volcanic plug Os Dois Irmãos on the coast, and the Ilhéu Caroço (Boné de Jóquei) in the distance (Peter Jones)

Plate 28 Ilhéu Caroço (= Boné de Jóquei, the Jockey's Cap), southeastern Príncipe, home of the endemic race of the Príncipe Seedeater *Poliospiza rufobrunneus fradei* (Jorge Palmeirim)

Plate 29 The crater lake Lago A Pot and the Pico do Fogo (435 m), Annobón (Jaime Pérez del Val)

Plate 30 The northwest coast of Annobón (Jaime Pérez del Val)

Plate 31 Tinhosa Grande (Ilhas Tinhosas), 25 km south of Príncipe (Martim Melo)

Plate 32 Nesting colony of Sooty Terns *Sterna fuscata*, Tinhosa Grande (Martim Melo)

Plate 33   The Sete Pedras, *c* 5 km off the southeastern coast of São Tomé (Peter Jones)

Plate 34   The Sete Pedras are the most important seabird breeding colony off the shore of São Tomé (Phil Atkinson)

Black Noddy   *Anous minutus*   BS;(B?)S;BS
                                   Boie 1844

A: Iguéle (B 1893b); Igualé, Iñelé (probable misprint for Igualé) (Bar); Íguel (bas);
Nodi de caperuza blanca (Sp: Bas). Port: Gaviota-de-bico-delgado (but refers to *A.
tenuirostris*) (F&V); Garajau-de-cabeça-branca (N 1994, C&C).

Tropical and subtropical Atlantic and Pacific; *A. m. atlanticus* in Africa; syn. *A.
leucocapillus* (Bocage 1903). Breeds Príncipe and Annobón, unrecorded São Tomé.

Príncipe Several collected at the islets of Pedra da Galé and Pedras Tinhosas, Nov and
Mar, including a juvenile on Pedras Tinhosas in Mar (Frade & Vieira dos Santos 1977)
and seen at Galé and Tinhosas in Aug (Sargeant 1994, Christy 1995b). Probably does
not breed on the tiny, wave-washed Galé (Naurois 1973a) but still breeds on Pedras
Tinhosas, although numbers (5000 pairs) given by Britton (*in* Urban *et al* 1986) have
not been critically determined (R. de Naurois *per* P.L. Britton *in litt*); 20,000–50,000
pairs according to Naurois (1973a).

São Tomé Reported on Sete Pedras, Mar (Christy & Clarke 1998): the only record for
the island.

Annobón First mentioned by Barrena (1911). Up to 14,000 pairs bred on sea-cliffs of
the main island and Tortuga islet in the 1950s (Fry 1961); 100 on Tortuga, May 1964 or
1965 (Robins 1966). The source of the figure of 70,000 pairs quoted by Urban *et al*
(1986) is not known. J. Pérez del Val (2001 and *in litt*.) suspects that Fry's figure was an
overestimate; he found it common in all suitable places, but not in the numbers
recorded by Fry. Probably the commonest seabird on the southern islets of Adams,
Santarém, Escobár and Fernando Póo, with several thousand birds present, May
(Robins 1966). Earlier, also recorded on the main island, nesting in a cavern in the
mountains, by Allen & Thomson (1848) and considered common on Tortuga by
Newton (Bocage 1893b) and very abundant on the 'less accessible, rocky parts of the
coast' by Fea (Salvadori 1903c).

Habits Mostly fishes inshore, flying low, and often in the company of Common
Noddy *A. stolidus* (Naurois 1994), in contrast to the sea-going Sooty Tern *Sterna
fuscata*, this perhaps accounting for the smaller numbers of noddies relative to the
latter species (Naurois 1973a).

Breeding The nest was described by Basilio (1957), Fry (1961), Naurois (1973a, 1994)
and Monteiro *et al* (1997), and eggs by Naurois (1973a) and Monteiro *et al* (1997).
Nesting apparently begins Jul but the main season is after Aug (Fry 1961); on Tinhosa
Grande, 1997, only adults present Apr, most adults with eggs Jun, most with chicks
Jul (Monteiro *et al* 1997); adults with eggs and young chicks, Sep–Oct 1997 (P. Kaestner
in litt to P. Atkinson). Believed by Basilio (1957), who collected a female with small
gonads in Sep, to breed year-round. Reproductive cycle thought to be annual by
Naurois (1973a).

Common Noddy   *Anous stolidus*   BS;BS;BS
                                    (Linnaeus 1758)

STP: Andorinha-do-mar (K); Padé do mal (corruption of Pardal-do-mar) (B 1904,
F&V); Palé (Monteiro *et al* 1997). A: Coicoi (Bar, Bas); Nodi común (Sp: Bas). Port:
Garajau-pardo (N 1994, C&C).

Tropical and subtropical oceans; *A. s. stolidus* on W coasts of Africa. Breeds on all
three islands.

Príncipe Not recorded by Dohrn (1866) although Keulemans (1866) reported it there, but only off the S coast. Later collected at the islets of Pedra da Galé and Pedras Tinhosas, Nov and Mar (Frade & Vieira dos Santos 1977) and seen at Galé and Tinhosas, Aug (Sargeant 1994, Christy 1995b). Breeds on Pedras Tinhosas, with at least 10,000 birds present (Britton *in* Urban *et al* 1986, R. de Naurois *per* P.L. Britton *in litt*), perhaps 20,000–50,000 pairs (Naurois 1973a); but conditions on the tiny, wave-washed Galé would probably rarely allow a nest to reach full term (Naurois 1973a).

São Tomé Keulemans (1866) reported it as more common here than on Príncipe. Moller collected one in 1885 (Bocage 1887c, Themido 1938), Alexander another (Bannerman 1915a), and Newton collected it, together with eggs, on the Sete Pedras rocks and on Ilhéu das Rolas, on both of which it probably still breeds (Bocage 1891, 1904, Monteiro *et al* 1997). Up to 200 on Sete Pedras, Mar (Christy & Clarke 1998).

Annobón About 1500 pairs bred on sea-cliffs of the main island and Tortuga islet in the 1950s (Fry 1961); *c* 10 birds were on Tortuga in May 1964 or 1965 (Robins 1966). Formerly recorded as common on Tortuga by Newton (Bocage 1893b); also collected by Fea, who considered it less common than *A. minutus* in 1902 (Salvadori 1903c). More common on the southern islets of Adams, Santarém, Escobár and Fernando Póo, with several thousand present in May (Robins 1966). Reported at sea at *c* 2°1'S 7°39'E, 12 Nov 1979 (Cadée 1981).

Habits Mostly feeds not at great distances from land, in contrast to Sooty Tern *Sterna fuscata*, this perhaps accounting for the smaller numbers of noddies relative to the latter species (Naurois 1973a).

Breeding On Pedras Tinhosas, nests in mixed colonies with Sooty Tern (Naurois 1994). The egg was described by Bocage (1891) and Naurois (1973a), nest-sites by Basilio (1957), Fry (1961), Naurois (1973a, 1994) and Monteiro *et al* (1997). Laying recorded Jun, Aug, Sep (Naurois 1994) but the main season is later than Aug (Fry 1961). Fledglings on Sete Pedras Jan 1997 but only adults there (including moulting birds) Mar and Jul; adult with fledgling on Rolas, Jul (Monteiro *et al* 1997). On Tinhosa Grande, most incubating eggs, Jan 1996; fledglings, Apr, but only adults (including moulting birds) there Jul; few carrying nest material or incubating, Aug 1995 (Monteiro *et al* 1997, Christy & Clarke 1998). Believed by Basilio (1957) to breed year-round but cycle thought annual by Naurois (1973a).

COLUMBIDAE

| **African Green Pigeon** | *Treron calva virescens* | **RB*;–;–** |
| | Amadon 1953 | |

Cessa (K, R); Ceci (Sousa 1887); Cécia, Cessía (B 1903, Correia 1928–29a, N 1972b, C&C); Rolla branca (K: name of partially albinistic variety). Port: Pombo-verde (F&V, N 1994, C&C).

Subspecies endemic to Príncipe, others on Bioko and African mainland. Forms a superspecies with *T. australis*, *T. sanctithomae* and *T. pembaensis* (treated as conspecific by Dowsett & Forbes-Watson 1993). The reference to *T. calva virescens* on São Tomé in Naurois (1988a) is an error. An albinistic form was apparently fairly common in the mid-19th century, but has never been noted since (Keulemans 1866, Naurois 1988a). Common to abundant in dense forest (Naurois 1983a), along edges of forest and regrowth, and in shaded cocoa plantations, in groups of up to 6 (Jones & Tye 1988, Naurois 1988a, Christy & Clarke 1998), mainly between 100 and 400 m altitude (Naurois 1988a). Apparently much more abundant in the 19th century, when

Keulemans (1866) found it the commonest of the three pigeons then known on Príncipe; occurring in groups of up to 50 birds. A favoured quarry species (eg Keulemans 1866) and still hunted for food (Jones & Tye 1988), so perhaps reduced by hunting since the 19th century (Naurois 1988a). Hunting pressure is probably responsible for its extreme shyness (Keulemans 1866, Naurois 1994).

Habits The call was briefly described by Keulemans (1866) and Christy & Clarke (1998); song heard frequently, August (Jones & Tye 1988), closely resembled that of southern African birds. Usually found in tree-tops (Snow 1950, Naurois 1988a, J&T), especially in fruiting trees, where they sit motionless during the hot part of the day (Keulemans 1866). Feeds on small fruit, including berries and (especially) figs *Ficus* spp and *Musanga*, and maize (Keulemans 1866, Naurois 1994, Christy & Clarke 1998).

Breeding Nest and egg described by Naurois (1988a, 1994). According to Dohrn (1866), eggs hatch Sep. Correia (1928–29a) collected a male incubating one egg in Jan, Naurois (1994) found a fresh egg in Mar. Fledgling Sep (Christy & Clarke 1998). Immatures collected Feb, May (Salvadori 1903a, Bannerman 1914). Keulemans (1866) thought them paired the whole year round, with a poorly defined breeding season; sparse data on state of gonad development and moult in specimens collected by Correia and Naurois suggested to Naurois (1988a, 1994) that the breeding season is Oct–Jun or Jul, centred on Mar, but perhaps all year; Christy & Clarke (1998) suggest the breeding season is Sep–Mar. Naurois (1988a) reviews data on the breeding season of the continental and Bioko subspecies.

São Tomé Green Pigeon   *Treron sanctithomae*   –;RB**;–
(Gmelin 1789)

Cécia (B 1891, 1904, Correia 1928–29b, F&V); Cessa (R); Céssia (C&C). Port: Pombo-verde de São Tomé (N 1994).

Endemic to São Tomé; forms a superspecies with *T. australis, T. calva* and *T. pembaensis* (treated as conspecific by Dowsett & Forbes-Watson 1993).

Perhaps first recorded about 1500, by V. Fernandes ('Pombas ou seixes muytas'), based on a report by G. Piriz, although 'Seixes' were elsewhere listed by Fernandes as domestic birds (Monod *et al* 1951). At least formerly, abundant on Ilhéu das Rolas (Allen & Thomson 1848, Bocage 1891, 1904) but now probably extinct there (Atkinson *et al* 1991). Common on main island in forest, second-growth and forest-edge habitats, shaded coffee and cocoa plantations, in pairs, or in groups of up to 10 birds at fruiting trees (Jones & Tye 1988, Naurois 1988a, Atkinson *et al* 1991). Not recorded in the N and NE since the 19th century (Bocage 1904). Occurs from sea-level to at least 1600 m, where abundant (Harrison & Steele 1989), although Naurois (1988a) considered it commonest between 300 and 1400 m. Commonly hunted for food (Günther & Feiler 1985, Jones & Tye 1988, Atkinson *et al* 1991).

Habits Eats bananas (Correia 1928–29b) and attracted to *Ficus* spp, *Musanga, Pycnanthus, Phyllanthus discoideus, Aidia quintasii, Pauridiantha floribunda, Cecropia peltata* and other fruiting trees; usually found high in trees (Jones & Tye 1988, Naurois 1988a, Nadler 1993, Christy & Clarke 1998). In Jan ate the small fruit of a *Capparis* sp (Allen & Thomson 1848). Song similar to, but simpler than, mainland *T. calva*: a crooning rattle, with clucking sounds, eg *poooo, peeyoo, prreeyoo, pripup, poorr, perrr, perrr...* or *poorrr-perrr, chup*; tape-recorded (J&T, J&B, P.D. Alexander-Marrack); other descriptions in Christy & Clarke (1998). Song heard frequently in Jul and Aug (Jones & Tye 1988). Birds in fruiting trees were frequently seen chasing and displacing one another in Jul (J&T). Also several in a group gave a tail-wagging display: up (hesitate) down (hesitate) at the rate of two complete cycles per sec. The tail was spread and folded at the same time but this was not synchronised with the vertical movements (J&T). The display was interspersed by the birds displacing one another.

Breeding The egg described by Bocage (1891) was probably not of this species (Naurois 1988a). A nest and egg were described by Amadon (1953) and others by Naurois (1988a, 1994) and Harrison & Steele (1989); the nests were found in Jan, Feb and Nov. Birds with enlarged gonads have been collected Dec, Jan, Feb and May. These data suggest a breeding season of Dec–Apr, corresponding with the main fruiting season (Naurois 1988a).

São Tomé Bronze-naped Pigeon *Columba malherbii*   ** RB;RB;RB
Verreaux & Verreaux 1851

STP: Rola (Rola própria (N 1972b) is a misunderstanding). P: Pombo (K); Rolla (Correia 1928–29b). ST: Lôla (corruption of Rôla: B 1888c *et al*, F&V). A: Lola Esalibavan (S 1903c); Lol, Lola (Bar, Bas); Lola sasá (Bas: to distinguish from *C. larvata*); Paloma irisada de Malherbe (Sp: Bas).Port: Pombo de Malherbe (N 1994); Pombo-de-nuca-bronzeada (C&C).

Endemic to Príncipe, São Tomé and Annobón. Member of a superspecies with *C. delegorguei* and *C. iriditorques* of mainland Africa (treated as conspecific by Dowsett & Forbes-Watson 1993). The reference in Amadon (1953: p. 413, line 1) to *C. malherbii* on Fernando Po (Bioko) is a typographical error (*cf* lines 4–5). The pigeon described by Keulemans (1866) as '*Columba* ....?' was, by its description, this species.

Príncipe Frequent to common in N in forest and plantations with tall trees (Jones & Tye 1988, Atkinson *et al* 1991). Its present distribution in the S is unclear but Keulemans (1866) recorded it from the higher areas, especially in forest, and Christy and Clarke (1998) consider it less common in the SW than in the N. Has apparently always been less common on Príncipe than on São Tomé (eg Dohrn 1866, Keulemans 1866), although thought more common on Príncipe in the 1960s and 1970s (Naurois 1983a).

São Tomé Frequent to common in W, centre and S in similar habitat (Jones & Tye 1988); recorded in N by Newton (Bocage 1889c, 1891), and still occurs in savanna bush near Guadalupe (Günther & Feiler 1985) and Morro Peixe (Atkinson *et al* 1991) and in mangroves on the north coast (Christy & Clarke 1998). Up to 1600 m in primary and secondary forest and coffee and cocoa plantations in the north–centre (Günther & Feiler 1985, Sargeant 1994, Christy & Clarke 1998). Also occurs in coconut plantations in the south (Christy & Clarke 1998). Formerly occurred on Ilhéu das Rolas (Frade & Vieira dos Santos 1977) but now apparently extinct there (Atkinson *et al* 1991).

Annobón A dove was recorded on the island in the early years of the 16th century by Valentim Fernandes, who described it as so tame as to be caught or killed by hand or with a stick (Monod *et al* 1951); perhaps both this and *C. larvata* were tame just after the discovery and initial settlement of the island. Recorded near the lake and peak in 1892 (Bocage 1903) and Fea regarded it as very common in thick forest at 400–500 m in 1902 (Salvadori 1903c). Basilio (1957) saw several in 1955, and regarded it as mainly inhabiting the forest interior; Fry (1961) thought it not common in 1959, and Harrison (1990) saw only 2 birds which might have been this species, during a one-day visit in 1989. Pérez del Val (2001) found it common but with restricted habitat in 2000.

Habits Usually seen high in trees, in groups of up to 24 (Basilio 1957, Eccles 1988, Jones & Tye 1988) but Günther & Feiler (1985), Eccles (1988) and P.D. Alexander-Marrack (*in litt*) reported it foraging on the ground. Food plants include *Tetrorchidium didymostemon*, *Phyllanthus discoideus*, *Fagara macrophylla* and *Alchornea cordifolia* (Christy & Clarke 1998). Food on Annobón is fruits (Basilio 1957). Barrena (1911) lists fruit eaten by his 'Lol', but that name could refer to either this species or *Columba larvata*. Voice on Annobón reportedly guttural, monotonous (Salvadori 1903c). This may refer to a mating call, described for Príncipe birds as a rattling *irrit, irt*, etc by Keulemans (1866). Song on Príncipe reportedly like that of European Turtle Dove *Streptopelia turtur* (Keulemans 1866); on São Tomé gives a call *hoo, hoo, hoo, hoo, hoo*, all on the same pitch, lasting *c* 5 sec (J&T); a similar and other calls were described by Christy & Clarke (1998). Is hunted for food (Keulemans 1866, Atkinson *et al* 1991).

Breeding Nests found (on São Tomé?) in shaded cocoa plantations in valley bottoms near the coast by Naurois (1994), who described nest and eggs. Said to have been breeding on São Tomé, Feb (Alexander *in* Bannerman 1915a). A juvenile, collected Jun, was in full moult (Günther & Feiler 1985). Said to breed mainly Oct–Jan, occasionally later, on Príncipe (Keulemans 1866); a partly grown juvenile collected May (Salvadori 1903a). On São Tomé and Príncipe, nest-building Jan, nestlings Jan–Mar (Christy & Clarke 1998). A male and female collected on Annobón, Oct, had small gonads, while 4 males and a female collected Nov had much-enlarged gonads, indicating the start of breeding (Basilio 1957). A juvenile was collected on Annobón, Feb (Bannerman 1915b).

Lemon Dove	*Columba larvata*	**RB***
	(Temminck 1810)	

Port: Pomba-limão (F&V), Pomba-preta (C&C).

Three endemic subspecies, one on each of the three islands. Other subspecies on Bioko and African mainland.

	Columba larvata principalis	**RB*;–;–**
	(Hartlaub 1866)	

Rolla (K); Moké (Correia 1928–29b); Muquê (F&V); Monquem, Rola-preta (N 1972b); Lola (N&CA).

Subspecies endemic to Príncipe. The bird discussed by Keulemans (1866) as '*Columba (Turtur)* ...?' was, by its description, this species.
     Common to abundant in cocoa plantations with many shade trees and a heavy canopy, in dense secondary growth (Jones & Tye 1988, *cf* Alexander *in* Bannerman 1914), and in primary hill forest (Keulemans 1866, Snow 1950, Christy & Clarke 1998). Unlike the São Tomé race, additionally seen in the open on dirt roads, alongside Laughing Doves *Streptopelia senegalensis* (Snow 1950, Jones & Tye 1988).

Habits This and the São Tomé race both seen most frequently when foraging on the ground, when they walk in erratic circles, or if flushed, when they normally settle close by (*c* 5 m) on a low branch along which they pace nervously, peering at the observer (Jones & Tye 1988). Food includes snails (Christy & Clarke), and blue berries, on which they become very fat in May and Jun; the berries stain both their feathers and body fat (Keulemans 1866). Hunted for food (Keulemans 1866). Song described by Christy & Clarke (1998).

Breeding Nest-sites, nests and eggs described by Naurois (1994), who gave breeding season (without data) as Sep–Jan (Nov–Mar according to Naurois 1972b). A juvenile was collected May (Salvadori 1903a). Keulemans (1866) thought the breeding season ill-defined, and found young in Aug and Jan.

Columba larvata simplex   –;RB*;–
(Hartlaub 1849)

Munqué, Muquê (B 1888c *et al*, F&V); Rola preta, Rola escura (N 1972b); Muncanha (R); Rola, Muncanha (J&T); Mucanha (M).

Subspecies endemic to São Tomé. Variable in appearance, some individuals being indistinguishable from *C. l. principalis* and others like mainland races (Amadon 1953).

Common to abundant, singly or in pairs, on and near ground in primary forest, regrowth, coffee and cocoa plantations with shade trees, wherever there is dense cover (Jones & Tye 1988) from sea-level up to the peak (Naurois 1988a, J&T, Harrison & Steele 1989, Christy & Clarke 1998). Mainly in southern and central forested areas but also recorded near Guadalupe and Praia das Conchas (N) (Bocage 1891, 1904, Günther & Feiler 1985, Sargeant 1994); Jones & Tye (1988) recorded 2 birds, probably of this species, in a small patch of dense, closed-canopy, savanna woodland near Morro Peixe (N), where also seen by Atkinson *et al* (1994); in the north, mainly in gallery-type forests (Christy & Clarke 1998). Newton also found it in the N and on Ilhéu das Rolas (Bocage 1891, 1904), and Alexander recorded it in palm groves and 'thick osier-like beds of the streams' near the capital (Bannerman 1915a); Nadler (1993) saw a group of 6 flying over Rolas in Apr 1991.

Habits Song a liquid *wup, wup, wup, wup, wup, wup, wup, wup, wup, wup, wup, wup*; 12–24 *wups* in a song, at the rate of 2–2.5 per sec; tape-recorded (J&B, P.D. Alexander-Marrack); calls also described by Christy & Clarke (1998).

Breeding Nest-sites, nests and eggs described by Bocage (1891, 1904) and Naurois (1994). Others were attributed to this species by Amadon (1953) but, from the description of the bird (Correia's 1928–29b 'Gray doves lives in town and at Rocas but around houses its use to feed with tame pigeons and hens'), nests (very flimsy structures of tendrils and stems), nest-sites (4 and 6 feet above ground in small 'mulberry' trees in a grassy field) and egg measurements (average of 4: 26.4 x 20.9 mm), these probably belonged to Laughing Dove *Streptopelia senegalensis* (*qv*). Also, Correia (1928–29a) used 'Ground Dove', not 'Gray Dove', to refer to *C. larvata*. Alexander found it breeding in early Feb (Bannerman 1915a). A juvenile was collected Aug (Salvadori 1903b); Naurois (1994) gave breeding season as Sep–Feb, but it is not clear on what evidence.

Morphometrics of a trapped individual were given in Atkinson *et al* (1994).

Columba larvata hypoleuca   –;–;RB*
(Salvadori 1903c)

Lola san-san (S 1903c but Bas refers this name to *C. malherbii*); Lol, Lola (Bar, Bas). Sp: Tórtola de Annobón (Bas).

Subspecies endemic to Annobón (subsumed in *C. l. inornata* of Bioko and W Africa by Urban *et al* 1986).

A dove was recorded on the island in the early years of the 16th century by Valentim Fernandes, who described it as so tame as to be caught or killed by hand or with a stick (Monod *et al* 1951); perhaps both this and *C. malherbii* were tame just after the discovery and initial settlement of the island. First collected by Fea, who considered it 'rare' in 1902 (Salvadori 1903c); not found by Alexander (Bannerman 1915b). Apparently common in the 1950s (Basilio 1957, Fry 1961), but Harrison (1990) failed to find it during one day in 1989.

Habits Forages on the ground, where commonly caught in traps by local people (Basilio 1957). Barrena (1911) lists fruit eaten by his 'Lol', but that name could refer to either this species or *Columba malherbii*.

Maroon Pigeon	*Columba thomensis*	–;RB**;–
	Bocage 1888d	

Pombo bravo, Pombo marreta (both names vernacular = Port: B 1888d, 1904, F, N 1994); Pomba pintada (N 1972b); Pomba preta (N 1988a); Pomba (R); Pombo, Pombo-do-mato (C&C).

Endemic to São Tomé. Forms a superspecies with *C. sjostedti* of Bioko and Cameroon and *C. arquatrix* of mainland Africa (treated as conspecific by Dowsett & Forbes-Watson 1993) but *C. thomensis* resembles juveniles of the latter species (Naurois 1994). Has an unusually long tail (Günther & Feiler 1985, J&T).

Generally common in forests and second growth at high altitude in the central massif (Correia *in* Amadon 1953, Snow 1950, Frade 1959, Günther & Feiler 1985, Naurois 1988a, Harrison & Steele 1989, Atkinson *et al* 1991, J&B), eg at Lagoa Amélia (1400 m), Calvário (1600 m) and near the Pico (up to 2000 m). There are also recent records from cultivated areas near Lagoa Amélia (Alexander-Marrack 1990, Atkinson *et al* 1991). Note that the 'locality' Capoeira quoted by Collar & Stuart (1985) from Frade ('1957': actual publication date 1959) is a vegetation type, not a locality. Frade (1959) stated: 'sometimes seen in high places where plantations have been abandoned, and a secondary forest has established itself (Capoeira)'. *Capoeira* is a Portuguese word for such forest. Also occurs in the isolated Pico Maria Fernandes, north of São João dos Angolares (Christy & Clarke 1998).

Probably a naturally montane species (see Amadon 1953) but not entirely restricted to high altitudes. Records below 1000 m include: Angolares (100 m, E, the type: Bocage 1888d); Ribeira Peixe (<100 m, SE) (Bocage 1904); Roça Granja (*c* 250 m, SE); Monte Café (*c* 800 m, N–centre) (Frade 1958); a juvenile male near sea-level at the Rio Caué estuary (SE) in Jul; 3 probable immatures at 50 m near São Miguel (SW) (Naurois 1988a and *in* Collar & Stuart 1985), Zampalma Velha (700 m, N–centre) (Jones & Tye 1988) and above Nova Moca (*c* 800 m, N–centre) (Sargeant & Alexander-Marrack 1990); 2 immatures in bamboo forest between the Rio Quija and Rio Xufexufe (5 m, SW); 4 single immatures along the Rio Ana Chaves (300 m, S–centre) (Atkinson *et al* 1991, 1994); 4 above Rio Xufexufe (Sargeant 1994). Eight of these records were in forests below 500 m; many of these were of immatures and it is not clear whether the lowlands support a breeding population.

The species also occurred formerly on Ilhéu das Rolas (Allen & Thomson 1848 '*Columba trigonigera*'; Bocage 1889b, 1891, 1904) which reaches only 96 m, but it is probably now extinct there (Naurois *in* Collar & Stuart 1985).

Although still fairly common at high altitudes, the main population of the Maroon Pigeon is restricted to a very small area of mid- to high-altitude primary forest. It is doubtful if the low-altitude areas are capable of supporting a viable population as the habitat is unsuitable or fragmented; several records there were of immature birds (see above) and their presence may be a result of post-juvenile dispersal. The species is very tame, and a favoured quarry with local people, who hunt it by waiting under fruiting trees, especially 'Pau cata' (Apocynaceae) or by lighting fires to which it is attracted to 'eat smoke' (smoke-bathe?) (Jones & Tye 1988, Christy 1995a). Despite considerable hunting pressure in the early 20th century (Correia 1928–29a), it still appeared fairly numerous in 1949 (Snow 1950) and the 1960s, though increased hunting had apparently made it rare by 1973 (Naurois 1983a, Collar & Stuart 1985). However, since independence, hunting has declined and at present levels is not considered a great threat, though any increase would give cause for concern (Jones & Tye 1988).

The Maroon Pigeon is termed 'Vulnerable, D1' (population <1000 adults) by BirdLife International (2000).

Habits The voice was described by Snow (1950), Naurois (1988a) and Christy & Clarke (1998). Four specimens had eaten berries of *Schefflera mannii*, which rarely occurs below 1400 m (Snow 1950, Monod 1960) but also reported to take fruit of cinnamon *Cinnamomum zeylanicum*, 'Pau cabra' *Trema guineense*, 'Pau ferro' *Margaritoria discoidea*, 'Pau matri', 'Pau cuidano' (both unidentified), and an *Alchornea* sp (pers comm by local guides), and observed in trees of *Canthium*, *Phyllanthus discoideus* and 'Nicolau' *Pauridiantha floribunda* (Christy & Clarke 1998). May is reportedly the peak time for finding large flocks in fruiting trees (Harrison & Steele 1989).

Breeding Birds collected by Snow, Sep, had inactive gonads but those taken by Correia, late Jul and early Aug, had enlarged gonads (Collar & Stuart 1985). The young bird, collected in Jul at the Rio Caué by Naurois (1988a), was estimated to have hatched in May. The nest and eggs are undescribed.

| **Feral Pigeon** | *Columba livia* | RB;RB;– |
| | Gmelin 1789 | |

Cosmopolitan (*C. l. domestica*). Resident São Tomé and Príncipe.

São Tomé and Príncipe Common to abundant in towns, villages and around plantation buildings (Eccles 1988, Jones & Tye 1988). Less common than the Laughing Dove *Streptopelia senegalensis* in the suburbs of São Tomé city (Jones & Tye 1988). Not mentioned by earlier ornithologists, although reported on São Tomé about 1500 by G. Piriz ('Pombos mãsos': Monod *et al* 1951).

| **Laughing Dove** | *Streptopelia senegalensis* | RB;RB;– |
| | (Linnaeus 1766) | |

Rola (F&V, N 1972b); Corucucú, Curucuco, Curucucú (N 1972b, R, C&C). Port: Rola do Senegal (N 1994, C&C).

Afrotropical and Palearctic. Resident São Tomé and Príncipe. The reference to it on Annobón by Naurois (1994) appears to be an error. We cannot find any difference between island and mainland specimens at BMNH (AT), and follow Amadon (1953) in not recognising the endemic race '*thome*' (see also Naurois 1988a).

São Tomé and Príncipe Domestic 'Rollas' (doves) were recorded on São Tomé about 1500 by G. Piriz, in addition to domestic pigeons or 'Pombos mãsos', presumably *C. livia* (Monod *et al* 1951); but it is not clear which species the 'Rollas' were. Otherwise, Laughing Doves were first recorded on São Tomé about 1900 (Bocage 1904) when they were evidently uncommon (Bannerman 1915a), and may have only recently colonised the island. First seen on Príncipe in 1949 by Snow (1950), who was told that it had been introduced about 1935 from São Tomé (D.W. Snow pers comm). Now common to abundant on São Tomé and common on Príncipe, along roads, in open plantations and in all open habitats with bare or nearly bare ground, including towns (Jones & Tye 1988). On São Tomé up to about 1100 m, eg at Bom Sucesso (Snow 1950, Christy & Clarke 1998) and 4 at Lagoa Amélia in Jun 1990 (1300 m) (Atkinson *et al* 1994); on Príncipe below 200 m (Naurois 1988a). Only locally common (eg at Porto Alegre) in S São Tomé, where its habitat is more restricted (Jones & Tye 1988). Common on Ilhéu das Rolas in April 1991 (Nadler 1993). On São Tomé apparently replaced by Lemon Dove *Columba larvata* on overgrown roads and tracks (Jones & Tye 1988). Naurois (1988a) found it most abundant in coconut plantations (presumably well-maintained ones) in N and NE São Tomé; does not occur in overgrown, neglected ones (J&T).

Habits Stomachs contained various seeds, including millet (Naurois 1988a); also eats fragments of coconut flesh (Christy & Clarke 1998). Calls as on mainland (J&T), described by Christy & Clarke (1998).

Breeding Nest and eggs described by Naurois (1988a, 1994), who found active nests in Nov and Dec. Amadon (1953) described 2 nests with eggs, collected by Correia on 2 Mar 1929, which he attributed to Lemon Dove *Columba larvata* but which were almost certainly of this species (see under *C. l. simplex*). Song heard often in Jul and Aug, and one bird seen carrying nest material (Jones & Tye 1988). A fledged juvenile incapable of full flight seen on São Tomé, 9 Jun (Günther & Feiler 1985). These records, and the gonadal state of specimens (Naurois 1988a) suggest a breeding season including Feb, May–Jul, Nov–Dec; may breed all year.

Morphometrics of a trapped bird given by Atkinson *et al* (1994).

PSITTACIDAE

Grey Parrot *Psittacus erithacus* RB;RB?;–
  Linnaeus 1758

Papagaio (do Príncipe) (K, F&V, R). Port: Papagaio-cinzento (N 1994, C&C).

Afrotropical including Bioko. Resident Príncipe; increasing frequency of reports from São Tomé. Following Amadon (1953) we do not recognise an endemic race *'princeps'*, having failed to find differences in a comparison of island birds and mainland *P. e. erithacus* at BMNH (AT, *cf* Benson *et al* in Fry *et al* 1988). Morphological information was presented by Melo (1998).

Príncipe Abundant since at least the mid-19th century (Dohrn 1866, Keulemans 1866), although Alexander considered that they had declined in numbers by 1909 (Bannerman 1914). However, it was abundant in the mid-20th century (Snow 1950, Naurois 1983a,b) although Naurois (1983a) reported another decline in 1968, perhaps due to pesticides. The population has apparently recovered since: in 1987 it was common to very abundant wherever there were tall fruiting trees (Jones & Tye 1988). Reported in the mountains by Keulemans (1866). In the evening most or all birds from the N of the island fly noisily, in groups of up to 30 birds, to roost on Pico Papagaio, and those from the S to Pico Negro, when they are very conspicuous (Dohrn 1866, Keulemans 1866, Correia 1928–29b, Snow 1950, Jones & Tye 1988). However, this may not occur during the breeding season, when birds are seen mainly in pairs, except when congregating at fruiting trees (Harrison & Steele 1989).

São Tomé One early report by Lopes de Lima (Hartlaub 1857). More recently, reported by local informants to Naurois (1972b, 1983b), who claimed that it bred in small numbers (2 or 3 small colonies) on the N coast; 3 birds at Ferreira Governo (N), Aug 1990 (Atkinson *et al* 1994); in Apr 1991, small groups flying south at dusk over Almerim and the city, one individual in a coconut palm on Ilhéu das Rolas and 5 flying over the islet (Nadler 1993). There is evidently a small, growing population on São Tomé, perhaps established by escaped cage birds or stragglers driven by storms (see Salvadori 1903b, F. Newton *in* Naurois 1983b), or it may have been an irregular visitor from the mainland (Naurois 1983b; see Splendid Glossy Starling *Lamprotornis splendidus* account). The highest densities were identified as around Ubabudo, Quimpo, Mestre António, Pinheira (with roost trees) and Mendes da Silva by Melo (1998), who also lists other localities. Newton repeated with scepticism the tale given by Dohrn (1866) and Keulemans (1866) that any parrot arriving on São Tomé would be killed by the Black (Yellow-billed) Kites *Milvus migrans* there.

Habits Eats fruit and seeds, especially of oil-palm *Elaeis guineensis* (Keulemans 1866, Naurois 1983b) and figs *Ficus* spp with large fruit, also *Fagara macrophylla* (Christy & Clarke 1998). Other foods and behavioural details were given by Melo (1998). Calls tape-recorded (J&T, J&B).

Breeding Said to breed on Príncipe in Dec (Keulemans 1866), and data available to Naurois (1983b, 1994) show egg-laying only in Nov, Dec and early Jan, but Alexander reported it 'breeding' in Mar (Bannerman 1914). Trappers regard the breeding season as Nov–Feb (Harrison & Steele 1989, Melo 1998). Eggs, nests, nest-sites, coloniality and other aspects of breeding behaviour were described by Keulemans (1866), Naurois (1983b) and Melo (1998). All nests found by Melo (1998) were *c* 20 m up in trees of *Funtumia africana* and 'Viro' (*Cleistanthus libericus* or *Scytopetalum klaineanum*). Reportedly aggressive to other species (Keulemans 1866; see under *Milvus migrans*).

Threats At least formerly (Keulemans 1866, Bannerman 1914, Frade 1958) hunted for food. Still caught in large numbers on Príncipe for export but not, apparently, on São Tomé: in the 1960s (despite protection in law) and since, chicks were taken from the nest and adults and juveniles caught (Naurois 1983b, Harrison & Steele 1989, Gascoigne 1995, Melo 1998). Naurois thought that only 10–20 birds were captured in this way each season, but numbers exported in 1987 and 1989 far exceeded that figure (J&T, Harrison & Steele 1989) and may reach 1000 per year (Melo 1998). This trade may decline if plans to regulate export are implemented (Jones & Tye 1988, Melo 1998). A method of capturing the adults was described by Keulemans (1866), who reported that nestlings were not taken in his day because of a belief that it was so hot inside the nest as to burn the hand!

Red-headed Lovebird   *Agapornis pullaria*   E(RB?);RB;–
                                                (Linnaeus 1758)

Periquito, Periquita, Peliquito (de São Tomé) (K, B 1891, F&V, R, N 1972b, 1994). Port: Periquito-de-bico-vermelho (C&C).

Afrotropical including (formerly?) Bioko; *A. p. pullaria* in W Africa. Extinct Príncipe, resident São Tomé, perhaps introduced.

Príncipe Reported once, by Keulemans (1866), who considered it uncommon in the uninhabited parts of the island; not seen by Dohrn (1866) in the same year, nor by anyone since. Finsch (1868) and Salvadori (1903a) considered it probably an irregular visitor or escapee. Such a conspicuous bird could scarcely have gone unnoticed for more than a century and we consider it extinct, if ever resident.

São Tomé Found by all early collectors and abundant in the late 19th century (Bocage 1891). Correia (1928–29b) found them much less common, probably because they were hunted. Two specimens from São Tomé at MNCN, prepared by P. Curats in 1943, were probably cage birds taken to Bioko (J. Pérez del Val *in litt*). Günther & Feiler (1985) saw few in 1983 and considered it less common than in the 1970s (*cf* Naurois 1983b). More recently, frequent to common in small flocks, throughout the lowlands, in open, scrubby habitats including gallery forest, forest edge, roadsides and open plantations (Jones & Tye 1988, Nadler 1993, Christy & Clarke 1998). Also recorded in forest edge, second growth and shaded plantations, at moderate altitude (Correia 1928–29b, Naurois 1983b, J&B). Five on Ilhéu das Rolas, Apr 1991 (Nadler 1993).

Habits Usually seen in twos and threes, occasionally up to 30 birds, flying, twittering, overhead; perches in tall trees (Correia 1928–29b, Naurois 1983b, Jones & Tye 1988). Calls tape-recorded (J&T). Food includes small seeds (Naurois 1983b) and plum-like fruit of 'Safu' (Correia 1928–29b).

Breeding The nest (in termitaria) and egg were described by Bocage (1904) and Naurois (1983b, 1994). The breeding season includes Oct–Dec and possibly also Jul (Naurois 1972b, 1983b), although Naurois (1994) stated (without presenting data) that laying occurs Nov–Dec and possibly Feb–Mar.

Has long been caught for export (eg Fry 1961, Eccles 1988, Harrison & Steele 1989) but such trade may eventually be banned (Jones & Tye 1988). Coimbra Museum contains an albino specimen from São Tomé (Themido 1938).

CUCULIDAE

Jacobin Cuckoo   *Oxylophus jacobinus*   VA;VA;–
                       (Boddaert 1783)

Port: Cuco-jacobino (C&C).

Africa and S Asia. Vagrant Príncipe and São Tomé.

Príncipe An immature in the middle Rio Papagaio valley, above Bela Vista, Jan (Christy & Clarke 1998).

São Tomé One specimen record, a juvenile moulting into adult plumage, collected on the beach near Roça Jou (SW) in Dec 1928 by Correia (1928–29b). It was probably of the race *pica* (Amadon 1953, S. Keith *in litt*), which breeds in the northern savannas of Africa and migrates to southern Africa for the non-breeding season. One was seen at Praia Melão (NE) on 26 Dec 1984 (Reinius 1985 and *in litt*).

Great Spotted Cuckoo   *Clamator glandarius*   ?;–;V
                             (Linnaeus 1758)

Africa and Middle East. One specimen apparently from Príncipe; vagrant Annobón, origin unknown.

Príncipe A specimen reputedly from Príncipe, but without other data, is in the Bocage Museum (Christy & Clarke 1998) or the Coimbra Museum (P. Christy *in litt*).

Annobón One record only, a young male, obtained on the shore by Newton between Nov 1892 and Jan 1893; Newton considered it accidental (Bocage 1903).

[European/African Cuckoo   *Cuculus canorus/gularis*   –;V;–

Palearctic migrant to Africa and Afrotropical resident. Vagrant São Tomé.

Recorded by Weiss, as described in an addendum by Hartlaub (1857, p. 266–267) in which he also mentions Weiss's record of São Tomé Oriole *Oriolus crassirostris*. Hartlaub's description does not make it clear to which of the two cukoo species the bird belonged; although it had small white tail-spots and the bill was stated to be blackish-horn, it was an immature (Hartlaub 1857).

Although neither the cuckoo nor the oriole was mentioned in Hartlaub's (1850) earlier description of Weiss's collection, the fact that they were dealt with together, and that most of Weiss's localities were reliable, suggests that this is a good record of one or other of these two cuckoos. Unfortunately, the specimen no longer exists in the Hamburg collection and was probably destroyed in World War II, as were the Museum catalogues (C. Hinkelmann *in litt*).]

Emerald Cuckoo   *Chrysococcyx cupreus insularum*   RB*;RB*;RB(ssp?)
Moreau & Chapin 1951

STP: Ossobó (B 1891 *et al, et auct.*); Pássaro-da-chuva (F&V); Bicho-brilhante (N 1972b); Chama-chuva (R). A: Fiájochi, Sebedol d'an (Sabedor de agua) (Bar); Filocotoy (Bar, Bas); Cuclillo dorado (Sp: Bas). Port: Cuco-esmeraldino (N 1994). Vernacular names onomatopoeic or referring to alleged habit of calling before or after rain.

Subspecies endemic to São Tomé and Príncipe; probably also this race on Annobón, where no specimen has ever been collected. Other subspecies on Bioko and African mainland.

Príncipe Much less common, at least during the 20th century, than on São Tomé (Correia 1928–29b, Naurois 1979). Not recorded by Jones & Tye (1988) though heard at Maria Correia and Rio Papagaio, Jul 1988 (J&B); considered common, Feb (Harrison & Steele 1989); frequent around Bom-bom and south of Santo António, Sep (Atkinson *et al* 1994); 2 heard, Aug (Sargeant 1994). Dohrn (1866) found it in the southern mountains 'during the dry season (from April to September)' and Keulemans (1866) considered it rare, and thought that it lived at higher altitudes in the 'summer' (= dry season) and lower in the 'rainy season'. Correia (1928–29b) made a similar comment, referring to both São Tomé and Príncipe. Alexander found it in Mar (Bannerman 1914) and Snow (1950) heard it in the N and centre in Sep.

São Tomé Apparently common in the late 19th and early 20th centuries (Bocage 1879, 1904). Although Bannerman (1915a) considered it rare at that time, he based this on his incorrect statement that none was collected between 1847 and 1909 (Bannerman 1915a, p. 107). Despite heavy collecting for its feathers, which reduced it near the city (Correia 1928–29a,b), it seems to have remained common away from human habitation, for Correia collected at least 59 (Correia 1928–29a). Persecution for its feathers continued until at least the 1950s. Numbers have increased since, except for a decline attributed to heavy pesticide use in the early 1970s (Naurois 1979; but described as 'abondant' at that time by Naurois 1983a). More recently, frequently heard in plantations in E, SE, SW and N–central Monte Café–Nova Moca area; common in W near Lembá and Bindá; also in the capital (Jones & Tye 1988, Harrison & Steele 1989, Nadler 1993, Atkinson *et al* 1994). Not found in dense regrowth of abandoned plantations, nor in dense primary forest above 1500 m by Naurois (1979) and Jones & Tye (1988). However, Harrison & Steele (1989) heard it commonly in Feb in primary forest up to 1600 m, and it was formerly common in higher altitude plantations (Snow 1950), which are now abandoned. Frequent near the Pico (Nadler 1993). Perhaps most abundant in cultivated cocoa plantations, with highest densities below about 200 m (Naurois 1979, J&T). Not heard in the drier N by Jones & Tye (1988) but others have recorded it there in dry woodland (Naurois 1979, Harrison & Steele 1989, Christy & Clarke 1998).

Annobón First noted by Barrena (1911); heard but not seen by Basilio (1957). Not found by Fry (1961) nor Harrison (1990) but Fry's visit occurred at a time when birds are mostly silent (calls most from Nov onwards: Basilio 1957) and Harrison had no time to visit the forested S of the island.

Habits Usually in dense tree crowns (Keulemans 1866, Basilio 1957), where difficult to see even when calling. The far-carrying call, well-described by Keulemans (1866), resembles that of the mainland race; tape-recorded on São Tomé (J&T, J&B). Both sexes sing (Keulemans 1866). Prey includes small beetles, snails and, especially, caterpillars (Keulemans 1907, Naurois 1979).

Breeding The egg is described by Bocage (1891, 1904) and Naurois (1979, 1994); it is apparently not mimetic. Keulemans (1866) was told by local people that hosts on Príncipe were Dohrn's Thrush-babbler *Horizorhinus dohrni* and Príncipe Speirops *Speirops leucophaeus* 'which breed in the summer', but thought that it also used other species. Breeding and other information given by Keulemans (1907). On São Tomé, the commonest host is probably São Tomé Weaver *Ploceus sanctithomae*; others include São Tomé (Newton's) Sunbird *Anabathmis newtonii* and São Tomé Prinia *Prinia molleri* (Bocage 1904, Naurois 1979, 1994). Potential hosts on Annobón include Annobón White-eye *Zosterops griseovirescens* and Annobón Paradise Flycatcher *Terpsiphone smithii*. The parent cuckoo apparently removes hosts' eggs (Naurois 1979).

The breeding season on Príncipe is unknown. On São Tomé it includes at least Oct–Dec (Naurois 1979) or Oct–Jan (Naurois 1972b). Naurois (1979) quotes 'negative' evidence, against breeding during the remainder of the year, which is not entirely convincing: although Correia found fewer birds with enlarged gonads May–Oct than Oct–Apr, he also obtained fewer birds in total for that period (Naurois 1979). Certainly, birds sing much Jul–Aug (Jones & Tye 1988) and the breeding season of known hosts extends Jul–Feb (Naurois 1979).

On São Tomé, is itself parasitised by its own endemic feather louse *Cuculoecus africanus*, the only one of this genus to have been found in the Afrotropical region (Mey 1993).

Klaas's Cuckoo   *Chrysococcyx klaas*   –;RB?;–
                           (Stephens 1815)

Sui-sui (Harrison & Steele 1989). Port: Cuco-bronzeado-menor (C&C).

Afrotropical including Bioko. Possibly resident São Tomé.

First reported when 2 birds were seen at forest edge and in second growth in the east, Apr 1987 (Eccles 1988). One seen in shaded plantation at 200 m, Feb 1989 (Harrison & Steele 1989). Its local name refers to its distinctive call.

TYTONIDAE

Barn Owl   *Tyto alba thomensis*   –;RB*;–
                     (Hartlaub 1852)

Cucúcú (B 1888a); Coruja (N 1972b, R). Port: Coruja de São Tomé, Coruja-das-torres, Coruja-branca (F&V, N 1994, C&C).

Cosmopolitan including Bioko. Subspecies endemic to São Tomé. No documented record on Príncipe (*contra* Wilson *et al* in Fry *et al* 1988).

Recorded along the whole of the E and N coasts, down the W coast as far as Bindá, and at Rio Quija, São Miguel and Porto Alegre (Bocage 1891, 1904, Salvadori 1903b, Amadon 1953, Jones & Tye 1988, Nadler 1993, Christy & Clarke 1998). Apparently only below 400 m (Naurois 1983b). Usually recorded near villages or roças; frequent in the city (Naurois 1983b, Günther & Feiler 1985), including in a church tower in the city (Naurois 1972b). Perhaps most common in poorly managed plantations and mixed woodland with oil-palm and *Erythrina* (Naurois 1983b). Formerly persecuted, because thought to take domestic chicks (Naurois & Castro Antunes 1973).

Habits Pellets contained almost exclusively rodent remains; once a passerine skull (Naurois 1972b, 1983b).

Breeding Nest-sites and eggs were described by Naurois (1983b, 1994). Available data indicate egg-laying extending over at least Dec–Jun (Amadon 1953, Naurois 1983b), although Naurois (1972b) thought it bred Jun to Sep or Oct; a pair with young, Aug (Christy & Clarke 1998).

STRIGIDAE

Annobón Scops Owl   *Otus scops feae*   –;–;RB*
                               (Salvadori 1903c)

Cucú (S 1903c); Cucuc (onomatopoeic; Bar, Bas), Autillo de Annobón (Sp: Bas); Mocho do Ano-Bom (Port: N 1994).

Palearctic and Afrotropical. Subspecies endemic to Annobón, others on African mainland.

Fea found it abundant in dense forest at 400–500 m in 1902 (Salvadori 1903c). Also found by Alexander (Bannerman 1915b) and Basilio (1957) but not by Fry (1961), nor since.

Reported by Fea to have a lightly trilled call at constant pitch, similar to that of the São Tomé Barn Owl *Tyto alba thomensis*, but higher-pitched; heard sometimes by day as well as at night (Salvadori 1903c). Stomachs contained insect and frog remains (Salvadori 1903c, Basilio 1957).

A male had enlarged gonads, Sep (Basilio 1957). Said to nest in tree holes (Basilio 1957). Breeding otherwise unknown.

[Príncipe scops] owl   *?Otus* sp   ?; –;–

Lobo (A. Gascoigne pers comm).

Príncipe Correia (1928–29a) and Melo (1998) recorded that local people told them about an owl that lived in the wild forests but was extremely rare. Also reported to A. Gascoigne (pers comm) in 1999, and there were sightings in the 1970s of an unidentified small owl (Naurois 1975a). These reports could refer to *O. hartlaubi, O. scops*, or another.

São Tomé Scops Owl   *Otus hartlaubi*   –;RB**;–
                                (Giebel 1872)

Cuco (B 1891); Coruja-pequena (Naurois 1975a); Kitóli (R); Quitóli (M). Port: Mocho-de-Hartlaub, Mocho de São Tomé (N 1994, C&C).

Endemic to São Tomé.

Uncommon; first recorded by Weiss in 1847 (Hartlaub 1850). Collected by Newton at São Miguel (SW), Angolares (SE), Roça Saudade (N–centre) and 'Roça Minho' (Bocage 1888d, 1891, 1904). Roça Minho does not appear on modern maps, but was unlikely to have been Roça (Dependência) Moinho (0°20'N 6°39'E), because Bocage (1891) gave the altitude of Minho as 1000 m, while Moinho is at only 285 m. The hill

Monte Minho (0°16′N 6°38′E) rises to 1062 m and it appears likely that Roça Minho may be a former name of the nearby Roça São Nicolau (950 m). In 1900 Fea found it at Ribeira Palma (NW) but, contrary to Naurois (1975a) and Collar & Stuart (1985), not at Água Izé in Dec (see Salvadori 1903b); Naurois seems to have confused Fea's records of this species and of *Tyto alba thomensis* (*Strix thomensis* of Salvadori 1903b) – Fea collected the latter at Água Izé in Dec. Correia collected one near Rio Ió Grande (Correia 1928–29a, Naurois 1975a), probably high in the hills near Cruzeiro (Correia 1928–29b). The species was not subsequently recorded for 45 years, until several were heard, seen or collected in 1973 and 1974 at Fortunato (c 400 m), Chamiço (c 900 m), Esperança (1300 m) and Lagoa Amélia (all N–centre), and near Rio Caué (S) (Naurois 1975a).

Less definite identifications were made by Snow (1950) at Zampalma (N–centre) and Jones & Tye (1988) at Ribeira Peixe (SE) in lowland secondary forest (not oil-palms, *contra* Atkinson *et al* 1991). More recently, the species was heard at Bom Sucesso, Jul 1988 (J&B), a pair was seen in Feb 1989 between Lagoa Amélia and Calvário (N–centre) (Harrison & Steele 1989), 4 birds were seen and tape-recorded between Nova Moca and Lagoa Amélia, Aug 1990 (P.D. Alexander-Marrack *in litt*). Local hunters reported that they knew the bird well by its call but seldom saw it (Harrison & Steele 1989). Atkinson *et al* (1991) found it widespread but never common in most habitats with tall trees, including primary and mature secondary forest up to 1600 m, but not in plantations, even those with shade trees. Night surveys found similar densities (average 2.6 birds calling per site) in montane rainforest (Lagoa Amélia), lowland secondary forest (Rio Quija) and lowland primary forest (Rios Xufexufe and Ana Chaves) (Atkinson *et al* 1991). In Mar–Apr 1991, it was heard at Cascata, Rio Angra Toldo and at 1100 m on the Pico (Nadler 1993).

These records suggest presence in forested country (including secondary forest) throughout the island and up to 1300 m. Most records come from secondary forest at moderate altitudes where Naurois (1975a) speculated that it was probably most common. It does not appear to have returned to low-altitude plantations, from which it was thought to have disappeared in the 1970s (Naurois 1975a). We agree with Naurois (1975a) that the species is probably not immediately threatened with extinction, but the population density is low enough to merit its classification as 'Vulnerable, D1' (population <1000 individuals) by BirdLife International (2000).

Habits Stomach contents included grasshoppers, beetles, a lepidopteran adult and bones (Naurois 1975a). The call was described as a single hollow whistle, lasting c 0.5s during which the pitch descends, with 10–15s between calls (tape-recorded by P.D. Alexander-Marrack and J.C. Sinclair; sonograms in Dowsett & Dowsett-Lemaire 1993; but see sonogram and descriptions in Nadler 1993 and Christy & Clarke 1998 for different durations and intervals). A different call was described as a soft, frog-like *crerr* or a growling *urrrr* (Harrison & Steele 1989, Atkinson *et al* 1994). Often calls during the day (J&B, P.D. Alexander-Marrack *in litt*).

Breeding Specimens taken in Apr, May, Oct and Nov had inactive gonads; a recent fledgling in Oct, and nestlings in Nov, indicate a breeding period Aug–Oct, preceding the season of heaviest rains; similar timing occurs in African populations of *O. scops* (see Fry *et al* 1988). Naurois (1975a) reasoned that Sep–Feb was the likely breeding season, since maximum rainfall in Oct–Dec would offer 'no advantage to newly-independent young', but this supposition seems unwarranted. The nest and eggs are unknown, but the juvenile was described by Naurois (1975a).

Systematics. The geographical origin of this species is a matter of debate. The vestigial ear-tufts and lack of feathering on the rear of the tarsus might suggest links with a group of mainly Asiatic scops owls (Naurois 1975a, Marshall 1978). However, given the physical difficulties posed by an Asian origin, a more likely possibility is that these characteristics have tended to evolve independently in several lineages. Tarsal feathering is variable within a species (Naurois 1975a, see data in Marshall 1978), so perhaps this characteristic is evolutionarily labile. Naurois (1975a) also notes that it is difficult to place *Otus* spp in inter-related groups.

The call is similar to that of a Eurasian Scops Owl *O. scops*, but with a longer interval between calls (Naurois 1975a); in the latter feature it resembles the calls of several other SE Asian and Indian Ocean spp (Marshall 1978), all of which, however, are small-island birds, so again this could be parallel evolution. The main call is nothing like the growl of the SE Asian *O. magicus*, although the call described by Harrison & Steele 1989 could be described as a growl; *O. magicus* is the species that (in some races only) most closely resembles *O. hartlaubi* in its tarsal feathering pattern.

In summary, the relationships of *O. hartlaubi* are still obscure, a fact admitted by Marshall (1978) and which counters his assertion that it is related 'to nothing in Africa'. A detailed morphological comparison between *O. hartlaubi* and congeners was presented by Naurois (1975a).

APODIDAE

São Tomé Spinetail	*Zoonavena thomensis* (Hartert 1900)	**RB**;RB**;–**

Andorinha, Andolim (B 1891, R); Andorinha-do-mato (N&CA). Port: Ferreiro-espinhoso, Rabo-espinhoso de São Tomé (F&V, N 1994, C&C).

Endemic to São Tomé and Príncipe. No longer considered closely related to Sabine's Spinetail *Rhaphidura sabini* of Bioko and mainland Africa (Fry *et al* 1988). The notes by Nadler (1993) on '*Chaetura sabini*' on São Tomé must refer to the present species.

Príncipe Correia (1928–29b) noted 2 species of swift on Príncipe but could not secure the smaller (which was presumably this species, as he reported recognising both Príncipe species from São Tomé). A spinetail was noticed in 1956 by Frade (1958) and identified as this species by Naurois (1985). It almost certainly breeds there, being frequent in E and NW in habitats similar to those occupied on São Tomé (Jones & Tye 1988). Frade & Vieira dos Santos (1977) reported that specimens from Príncipe are much smaller than those from São Tomé.

São Tomé Not obtained, but probably seen, by Newton; first collected by Mocquerys at Pedroma (Hartert 1900). Now frequent in SE, common in S, north–centre and W over open plantations, forest clearings and valleys (Jones & Tye 1988); also reported to forage inside plantations and open woodland, unlike other swifts (Naurois 1985). Up to at least 1300 m (Salvadori 1903b, Snow 1950, J&T, Christy & Clarke 1998) and perhaps to the summit of the Pico (Nadler 1993), and down to sea-level (J&T, Atkinson *et al* 1994, *contra* Naurois 1985). Recorded from northern savannas (Atkinson *et al* 1991).

Habits Often feeds within 5–10 m of ground, lower than other swifts (Jones & Tye 1988); perhaps higher in early afternoon (Naurois 1994). Frequently forms loose groups of up to 20 individuals, sometimes with other swifts (Günther & Feiler 1985, Atkinson *et al* 1991, J&T) on both islands (*contra* Naurois 1985, Fry *et al* 1988). Call a high-pitched *chip* (Sargeant 1994).

Breeding Seen entering holes in trees on Príncipe, possibly prospecting for nest-sites (Naurois 1985). Eggs, nestlings, immatures, nest and nest-sites on São Tomé described, and nest figured, by Naurois (1985); laying recorded Aug and Oct (Naurois 1994).

Palm Swift   *Cypsiurus parvus*   RB;RB;–
                                   (Lichtenstein 1823)

Ferreiro (F&V); Andorinha (R). Port: Ferreiro-das-palmeiras, Guincho-das-palmeiras (N 1994, C&C).

Afrotropical including Bioko, São Tomé and Príncipe; *C. p. brachypterus* on the islands. The population on Príncipe has been named as a subspecies *C. p. sanctosjuniori* (see Naurois 1994); if colonisation of the islands really took place only in the 20th century, this is surely misplaced.

São Tomé and Príncipe First recorded on Príncipe in 1949, when *c* 15 pairs were found breeding at a single site near Sundi, NW (Snow 1950), and on São Tomé in 1959, when it was common over the city in Jul (Fry 1961). Naurois (1983a) noted 'reproduction découverte', presumably for São Tomé, in 1972 (Naurois 1994 stated that it was in 1970, on São Tomé). Now well-established as a breeding bird on both islands, where it is abundant wherever there are palms (of any species) in fairly open habitat, including northern savannas, São Tomé City, coconut and oil-palm plantations, mountain valleys and mixed regrowth (Jones & Tye 1988). Not restricted to low altitude nor to savannas, nor to the vicinity of human habitation (*contra* Naurois 1983a, 1985), but perhaps more common in such habitats. On São Tomé, noted as uncommon at Rio Quija and at the summit of the Pico (Nadler 1993).

Breeding Nests described by Naurois (1972b, 1994). Naurois (1994) gave breeding season as Oct–Dec, without presenting data, whereas Naurois (1972b) gave breeding season on Príncipe as Nov–Mar. Fry *et al* (1988) stated that Príncipe birds breed Sep–Oct, but the source of this information is unknown (C.H. Fry *in litt*). Seen entering palm crowns on São Tomé, Jun–Aug (Günther & Feiler 1985, J&T), and probably breeding then. Apparently not breeding Mar–Apr (Nadler 1993).

[Black Swift   *Apus barbatus*   ?;?;–
                                   (Sclater 1865)

Africa to Madagascar; *A. b. sladeniae* Nigeria to Angola, including Bioko.

Príncipe: a possible sighting at the airport, Sep–Oct 1997 (P. Kaestner *in litt* to P. Atkinson).

São Tomé Observations of birds accompanying Little Swift *Apus affinis*, in the towns on São Tomé and Príncipe, and referred to this species (Christy & Clarke 1998), require confirmation.]

[Common Swift   *Apus apus*   –;?;–
                                   (Linnaeus 1758)

Port: Guincho-da-Europa.

Breeds Palearctic, winters Africa.

One record: 2 birds seen by P. Christy at Quija, São Tomé Apr 1994 (Christy & Clarke 1998, P. Christy *in litt*). This record requires confirmation.]

Little Swift   *Apus affinis bannermani*   RB*;RB*;–
                                     Hartert 1928

Pascusha (K); Andorinha, Andolim (B 1888a, 1891, R); Ferreiro (F&V); Andorinhão-pequeno (N 1994). Port: Guincho-pequeno (C&C).

Eurasia and Africa. Subspecies endemic to São Tomé and Príncipe. Mainland birds, including specimens from Cameroon, differ in several respects from the island race (Bannerman 1930–51 vol. 3, AT). Bioko specimens appear intermediate between those from São Tomé and those from the neighbouring mainland (S. Keith *in litt*, see also Amadon 1953). Subspecies on the African mainland include some individuals showing streaking on the throat, which is therefore not an exclusive character of the island birds (S. Keith *in litt*).

Príncipe Reported by Keulemans (1866) throughout the island, in 'plains', plantations, villages, mountains and 'in' the forests of the uninhabited parts. Dohrn (1866) found it common over the town. More recently, abundant over plantations, clearings and secondary growth; apparently generally more common than on São Tomé (Jones & Tye 1988). Breeds on the rocky peaks in the south (Christy & Clarke 1998).

São Tomé Abundant in the city and common in all other habitats, perhaps less so over dense forest, and not inside forest (Naurois 1985, Jones & Tye 1988), from sea-level to at least 850 m (J&T), and seen over the summit of the Pico by Nadler (1993).

Habits Can be seen in parties of up to 20, chasing and screaming in the evenings (Bannerman 1930–51 vol. 8, J&T). Flies from before sunrise until after sunset; rests hanging from tree trunks or nests (Keulemans 1866).

Breeding Nest-sites (on cliffs or in houses), nests and eggs described by Keulemans (1866), Snow (1950), Nadler (1993) and Naurois (1994). Visiting nest-sites on sea-cliffs near Morro Carregado (NW São Tomé) Jul–Aug (Jones & Tye 1988, Atkinson *et al* 1994); recorded nesting in São Tomé city and on a basalt cliff between Neves and Santa Catarina (NW) in Jun and Sep (Snow 1950, Günther & Feiler 1985); one nest with young Mar or Apr (Nadler 1993); Naurois (1972b) gave breeding season as Sep–Mar, but later stated that laying took place Dec–Jan (Naurois 1994). Breeding on Príncipe in Apr and May (Keulemans 1866) and visiting a nest-site there in Sep (Snow 1950). Colonies are occupied all year, and used as dormitories (Christy & Clarke 1998).

Alpine Swift   *Tachymarptis melba*   –;–;PM?
                                    (Linnaeus 1758)

Scüé, Scue (Bar, Bas); Vencejo real (Sp: Bas).

Breeds Eurasia and in isolated populations on African mainland, dispersing in non-breeding season. Perhaps regular Annobón.

First mentioned by Barrena (1911). Apparently a regular visitor, where reportedly well-known to local inhabitants; single birds seen several times, Oct–Nov 1955, hunting insects over the town and in a market garden (Basilio 1957). Basilio thought, by their timing, that the birds were most likely of the Eurasian migrant race *T. m. melba*. Basilio's description could not apply to any other African swift, nor to a pratincole or petrel.

ALCEDINIDAE

Blue-breasted Kingfisher   *Halcyon malimbica dryas*   RB*;–;–
Hartlaub 1854

Chocho (Dohrn 1866, R); Sho-sho (Correia 1928–29b); Chau-chau, Chó-chó (F&V, C&C) (onomatopoeic). Port: Pica-peixe-de-peito-azul (N 1994).

Subspecies endemic to Príncipe, others on African mainland. Three 19th-century records from São Tomé and others from Bioko are considered erroneous (Amadon 1953, Naurois 1980).

São Tomé One of the 3 specimens (BMNH 59.6.28.7, which is that referred to by Sharpe 1892b) was 'purchased of Verreaux' and another was recorded as 'St. Thomé: Gujon' (Hartlaub 1861), ie collected by Gujon. As localities given by both of these collectors were often unreliable (see Hartlaub 1852, 1857, 1861, Salvadori 1903b), these records are considered erroneous. The third São Tomé specimen, collected by Weiss, was recorded as 'Hab. Ilha do Principe: St. Thomé: Weiss' (Hartlaub 1853–54) and 'Ins. St. Thomé et do Principe: Weiss' (Hartlaub 1857). Several considerations suggest that this specimen may actually have come from Príncipe. First, Hartlaub (1850) reported the species only from Príncipe. Second, the form of words used by Hartlaub (1857) was that used by him when writing about other Príncipe specimens, eg Hartlaub (1850) said in 2 places that Príncipe Drongo *Dicrurus modestus* comes from Príncipe but in the formal description stated 'Habitat: ins. St. Thomé.'. Similarly, when referring to Príncipe Glossy Starling *Lamprotornis ornatus*, he stated 'Hab. Ilha do Principe: Weiss in Mus. Hamb. St. Thomé: Weiss ib.' (Hartlaub 1853–54) and 'Ilha do Principe; St. Thomé' (Hartlaub 1857). Since both the starling and the drongo are otherwise only known from Príncipe, these examples suggest that Hartlaub sometimes treated localities on the two islands loosely, and perhaps regarded Príncipe as part of, or a dependency of, the Portuguese province of São Tomé. However, later authors have assumed that one or more of Weiss's *Halcyon* specimens came from São Tomé (eg Bocage 1904, Bannerman 1915a). Dr C. Hinkelmann (*in litt*) has been able to track down 3 specimens of this subspecies collected by Weiss, in ZMH, although the relevant catalogues were destroyed in World War II. All bear the number RK17.248, with different Roman numbers. Nos. I and II are labelled as from Príncipe, while No. III (male) is labelled 'St. Thomé, coll. Dr. med. Weiß'. The labels were probably written in Europe, and Prof. W. Meise, who re-catalogued the surviving specimens after the war without having seen the original catalogues, noted of No. III 'ob Príncipe richtig? (vor 1847)' (rightly from Príncipe? (before 1847)). Contrary to Bannerman (1915a), Bocage (1889a, 1891, 1904) and Salvadori (1903b) stated clearly that Newton did not find it on São Tomé. We have been unable to trace any further records for São Tomé and conclude that there is no good evidence that it ever occurred there (*pace* Fry *et al* 1988).

Bioko The only record of *H. m. dryas* of which we are aware is the specimen BMNH 42.11.4.6, collected by Thomson (not by Fraser, *pace* Sharpe 1892b). Thomson's localities are also unreliable, probably owing to mixing of specimens during the chaos following the tragic Niger expedition (eg type locality for *Terpsiphone atrochalybeia* given as 'Fernando Po': Thomson 1842; see also Allen & Thomson 1848, Hartlaub 1852). Letters from Fraser in the Liverpool Museum indicate that Thomson was separated from the luggage tender for some time, and could easily have lost track of unlabelled specimens (A.M. Moore pers comm). Sharpe (1892b) referred to another specimen of *H. m. dryas* from Bioko, on the authority of Fraser, but this skin (BMNH 81.5.1.2988), from the Gould collection, was (and still is) unlabelled as to locality and Fraser's designation of Bioko appears to have been merely supposition. The species has not, therefore, been reliably recorded on Bioko, neither *dryas* nor any other race (J. Pérez del Val *in litt*).

Mainland Similarly, the only record, skin BMNH 74.10.1.4, from 'Eloby d. Gaboon' is probably also incorrectly located; it was collected by Ansell, many of whose localities seem erroneous.

The above evidence leads us to the conclusion that there have been no reliable records of *H. m. dryas* outside Príncipe.

Príncipe Appears to have recovered from decline attributed to pesticide use and hunting by Naurois & Castro Nunes (1973); seen and (especially) heard commonly in plantations, and in Santo António; also in forest and regrowth at all altitudes, though probably more common in the lowlands. Often but not always near streams (Keulemans 1866, Naurois 1980, Jones & Tye 1988). Recorded from Ilhéu Caroço (Naurois 1975b).

Habits Calls most often early morning and late afternoon (Keulemans 1866, Jones & Tye 1988) and reportedly calls on moonlit nights (Keulemans 1866). Calls and displays were described by Keulemans (1866), Snow (1950), Naurois (1980) and Christy & Clarke (1998); song and harsh, screeching call tape-recorded (J&T, J&B). Hunts mainly terrestrial prey, sometimes digging in the ground (Keulemans 1866, Naurois 1980). Prey includes snails, terrestrial and aquatic insects, including cerambycid beetles up to 7 cm long, earthworms, lizards, small fish, crustaceans and palm fruits (Dohrn 1866, Keulemans 1866, Snow 1950, Naurois 1980, Christy & Clarke 1998). At a snail 'anvil' site at first attributed to a Gulf of Guines Thrush but now believed to have been made by a Blue-breasted Kingfisher (J. Baillie and A. Gascoigne pers comm), the predominant prey (93%, n=81) was the endemic *Columna columna*, together with a few shells of another single-island endemic *Lignus* (formerly *Pseudotrochus*) *alabaster*, and juvenile *Archachatina bicarinata*, endemic to Príncipe and São Tomé (Christy & Gascoigne 1996). Takes nestlings, including Príncipe White-eye *Zosterops ficedulinus*, Dohrn's Thrush-babbler *Horizorhinus dohrni* and Príncipe Golden Weaver *Ploceus princeps* (Keulemans 1866, Correia 1928–29b) and reportedly captures young, free-flying birds, especially Bronze Mannikins *Lonchura cucullata* (Keulemans 1866). One captured a White-bellied Kingfisher *Corythornis leucogaster*, and others were seen to be mobbed by various passerines (Dohrn 1866, Snow 1950). One chased a fruit bat that had been taken out of a mist-net (M. Melo pers comm).

Breeding Nest and eggs described by Keulemans (1866) and Naurois (1972b, 1980, 1994). Nestlings recorded Oct, Nov and Feb (Frade 1958, Naurois 1980), which led Naurois (1994) to quote breeding season as Sep–Feb; red-billed juvenile seen Aug (Jones & Tye 1988); immatures collected Jan, Feb, Mar and May (Salvadori 1903a, Bannerman 1914).

Morphometrics of 2 trapped birds given by Atkinson *et al* (1994).

White-bellied Kingfisher　　　*Corythornis leucogaster nais*　　　**RB*;–;–**
　　　　　　　　　　　　　　　　(Kaup 1848)

Pica-peixe (Dohrn 1866, K, N 1994, C&C); Conóbio (N 1972b); Conóbia (C&C). Port: Guarda-rios do Príncipe (F&V); Pica-peixe-pequeno do Príncipe (N 1994); Pica-peixinho-de-barriga-branca (C&C).

Afrotropical. Subspecies endemic to Príncipe; others on Bioko and African mainland.
　　　Naurois (1980, Fry & Naurois 1985) considered this to be mainly a terrestrial bird, or (Naurois 1983a, 1994) not restricted to watercourses, occurring on drier

ground in woodland and more open vegetation on plateaux and slopes between streams. Others have found it often along the coast and near rivers in the lowlands, less frequently in the interior, though including gardens and plantations far from water (Dohrn 1866, Keulemans 1866, Snow 1950, Jones & Tye 1988, Sargeant 1994). We found it common in habitats similar to those occupied by the São Tomé species, including along the Rio Papagaio (J&T). Found throughout the island in all types of open and semi-open habitat (Christy & Clarke 1998).

Habits Food includes terrestrial and aquatic prey (*contra* Fry *et al* 1988), including shrimps (Jones & Tye 1988), fish and water insects (Keulemans 1866). Freshwater fish (perhaps one or other of the species listed for São Tomé, Bioko and Annobón by Thys van den Audenaerde 1967 and Zarske 1993) are present on Príncipe (J&T), contrary to information given to Snow (1950). Other prey include earthworms, spiders, blattids, grasshoppers, beetles, lizards and crabs; rejecta in two nests included lizard bones and remains of beetles, wasps, damselflies and crabs, and an adult was seen to feed a long lizard to a nestling (Snow 1950, Fry & Naurois 1985). Other behavioural notes were given by Keulemans (1866). Call and display described by Christy & Clarke (1998).

Breeding Nests by the sides of paths or on slopes or crests of ridges (Fry & Naurois 1985) or under tree roots (Keulemans 1866). Eggs white; clutch 2–3 (Naurois 1994) or up to 5 (Keulemans 1866); eggs described by Naurois (1994); male feeds incubating female. Reportedly breeds Aug–Jan (Keulemans 1866), or Sep–Feb (Naurois 1972b, 1994) or Dec–Jan (Fry *et al* 1988, although the origin of this latter statement is unknown: C.H. Fry *in litt*).

Morphology and systematics Raises crest when handled (Melo 1998), as does *C. cristata*. Morphometrics of 2 trapped birds given by Atkinson *et al* (1994).

Malachite Kingfisher   *Corythornis cristata thomensis*   –;RB*;–
                                  (Salvadori 1902)

Conóbia, Cunóbia (B 1879, 1891 *et al*, S 1903b, R); Conóbio, Conóbia, Pica-peixe (N 1972b, C&C). Port: Guarda-rios de São Tomé (F&V); Pica-peixe-pequeno de São Tomé (N 1994); Pica-peixinho-de-poupa (C&C).

Afrotropical. Subspecies endemic to São Tomé; related subspecies on mainland.

Common along streams and rivers, including mountain torrents, wherever there are shallow pools in which to hunt (Jones & Tye 1988); apparently territorial, pairs occupying stretches of *c* 300–500 m of stream (Nadler 1993, Atkinson *et al* 1994); sometimes found on the seashore, in estuaries or on lagoons (Snow 1950, Fry & Naurois 1985, Eccles 1988). Occurs at least up to about 1000 m (Snow 1950) but probably more common at lower altitudes, where its habitat is commoner (Fry & Naurois 1985). Occasionally seen away from water, where it hunts insects (J&T, Atkinson *et al* 1994, Christy & Clarke 1998). At least formerly occurred on Ilhéu das Rolas (Bocage 1891).

Habits Chattering and chipping calls tape-recorded (J&T). Captures small shrimps (Jones & Tye 1988); Snow (1950) saw one carrying a small fish; Naurois found probable fish remains below a nest (Fry & Naurois 1985) and thought they also ate insects (Naurois 1972b).

Breeding Nest excavated in stream bank or other earth bank (Fry & Naurois 1985), up to 600 m from water (Atkinson *et al* 1994). Eggs said to have been found May–Jul by Naurois (1994), who stated that it also lays Oct–Mar, whereas Naurois (1972b) gave breeding season as Dec–May. One nest with 4 young, Aug (Atkinson *et al* 1994); another with young, Apr (Christy & Clarke 1998). Many young birds (with all-black bills), as well as all other plumage states, seen in Jul and Aug (Salvadori 1903b, Jones & Tye 1988), and in Dec–Jan and Mar–Apr (Christy & Clarke 1998).

Morphology and systematics The immature of this form is very different in appearance from those of Príncipe and mainland forms (Salvadori 1902) and more closely resembles the Shining Blue Kingfisher *Alcedo quadribrachys*. Of taxonomic relevance is the fact that, when captured, the bird raises its crest, as does mainland *C. cristata* (J&B) and Príncipe *C. l. nais*. Morphometrics of trapped individuals were given in Atkinson *et al* (1994).

Corythornis spp

The taxonomic position of these two island forms of *Corythornis* kingfishers (*nais* on Príncipe and *thomensis* on São Tomé) is uncertain. Sibley & Monroe (1990) treat the two as endemic full species. They have been variously referred to the two mainland species *C. cristata* and *C. leucogaster* (see review in Fry & Naurois 1985) and are intermediate between them in plumage and some elements of structure (Amadon 1953, Fry & Naurois 1985). The suggestion of Fry & Naurois (1985), that the two island forms derived directly from the African mainland, is therefore more plausible than that of Amadon (1953), who suggested that the Príncipe birds derive from immigrants from Bioko and the São Tomé population from Príncipe.

However, we disagree on two points with Fry & Naurois (1985). First, they consider *nais* as derived from *leucogaster* rather than from *cristata*, because *leucogaster* and *nais* are less closely tied to aquatic habitats. However, our own and other observations of *nais* on Príncipe (see above) indicate that it is almost as much a waterside bird as *thomensis*, with most records of both in similar streamside habitat (Jones & Tye 1988, see also Keulemans 1866, Dohrn 1866). Second, Fry & Naurois (1985) reject the notion that *nais* and *thomensis* derived from invasions that occurred before the mainland forms separated into two good species, because they would 'expect a much greater distinction from either existing continental species to have evolved, in some proportion to the time which has elapsed'. Since evolutionary change does not proceed at a fixed rate, this conclusion does not necessarily follow. It is at least as likely that the two island forms represent invasions that occurred before the separation of the two mainland species, as it is that they represent later invasions, one of *leucogaster* and the other of *cristata*. On present evidence therefore, and until molecular data become available, it is not possible to determine satisfactorily the closest relatives of the island taxa. Unless they are given specific status, it is probably simplest to link them with the mainland taxa that they most closely resemble morphologically. This course produces *C. leucogaster nais* and *C. cristata thomensis* (see Fry & Naurois 1985), as we have adopted above. This is purely an arrangement of convenience.

| **Pied Kingfisher** | *Ceryle rudis* | ?(B);VA;– |
| | (Linnaeus 1758) | |

Chocho branco (Sousa 1887, B 1903). Port: Pica-peixe-malhado (C&C).

Afrotropical (including Bioko) and S Asia. Probably vagrant, São Tomé and Príncipe.

Príncipe One specimen collected by Newton at the mouth of the Rio Papagaio, Mar 1887 (Sousa 1887), when it apparently had a local name. Neither Dohrn nor Keulemans had found it. Not seen since, until S. d'Assis Lima (*in litt*) found a pair nesting at *c* 50 m altitude on the R. Papagaio between Santo António and Bela Vista in early June 1997. A pair seen in the bay at Santo António, Jun 1997 (Christy & Clarke 1998), were almost certainly the same birds.

São Tomé The only record is 2 birds near Praia Diogo Nunes (NE), Jan 1985 (Reinius 1985 and *in litt*).

MEROPIDAE

[Bee-eater   *Merops* sp   –;–;V

Annobón A bee-eater (species unknown) was seen in August 2000 (Pérez del Val 2001).]

CORACIIDAE

European Roller   *Coracias garrulus*   **VPM;VPM;–**
                                 Linnaeus 1758

Port: Rolieiro (F&V); Rolieiro da Europa (C&C).

Palearctic migrant to tropical Africa; *C. g. garrulus* in W Africa. Vagrant, São Tomé and Príncipe.

Príncipe Two birds seen along the shore in Nov and Dec 1865, of which one, a female, was shot; the other was probably a male (Keulemans 1866). One collected at the airport, Dec 1954 (Frade 1958). The birds seen by Keulemans hunted large insects, including grasshoppers. The species was not known to the inhabitants (Keulemans 1866).

São Tomé Two specimens are known, a young bird collected by Weiss in 1847 (Hartlaub 1850) and a female by Newton in 1891 (Bocage 1904). Weiss recorded it as 'completely unknown there' (Hartlaub 1850), presumably meaning to the local people. More recently, one seen over the northern grasslands, Feb 1989 (Harrison & Steele 1989).

HIRUNDINIDAE

Banded Martin   *Riparia cincta*   **VA;–;–**
                                  (Boddaert 1783)

Undurinha, Pascusha (K).

Afrotropical. Probably vagrant, Príncipe.

    Found by Dohrn (1866) and Keulemans (1866) each of whom collected a female (or perhaps only one specimen between them?). Keulemans recorded it in Jun, Sep and Oct, considering it uncommon, but thought it lived at higher altitude in the 'summer' (by which he meant the dry season) than in 'winter'. Both observers recorded it near the shore.

[Grey-rumped Swallow   *Pseudhirundo griseopyga*   **?;–;–**
                                  (Sundevall 1850)

Afrotropical.

Príncipe Reportedly seen at the airport, Sep–Oct 1997 by P. Kaestner (*in litt* to P. Atkinson).]

Barn Swallow   *Hirundo rustica*   **VPM;PM;–**
                                 Linnaeus 1758

Andorinha (F&V).

Palearctic migrant to Africa. Regular, at least on passage, São Tomé and Príncipe.

Príncipe A female was collected at the airport, Nov 1954 (Frade 1958); a bird ringed as a juvenile in Scotland, Aug 1974, was recovered on Príncipe, May 1975 (Mead & Clark 1987, C. J. Mead *in litt*).

São Tomé One was collected by Moller in 1885 (Bocage 1887c, Themido 1938); observed in the city by P. Alden and S. Reinius, Jan 1985 (Reinius 1985 and *in litt*); two at Quija lagoon, Mar 1994 (Christy & Clarke 1998); a small party seen flying north along the NE coast, Apr 1985 (M. Goulding *in litt*); a party of 3 foraging over Rio Quija, and 2 parties of 2 and 3 along shore between Rio Quija and Neves, 13–14 Apr 1991 (Nadler 1993).

House Martin   *Delichon urbicum*   **VPM;–;–**
                                 (Linnaeus 1758)

Palearctic migrant to Africa. Two records from Príncipe.
    One female was shot in a field at 450 m altitude, Jan 1865 (Keulemans 1966); 3 were seen near Santo António and Terreiro Velho, Sep 1928 (Correia 1928–29b, Amadon 1953).

MOTACILLIDAE

[Wagtail   *Motacilla* sp   **?;–;–**
Chu-tri-pa (K).

Perhaps vagrant Príncipe.
    Keulemans (1866) reported a wagtail as being fairly common, in groups of 5–7, in frequented places such as plantations. He described it as 'like the young of *M. alba*': possibly this species in winter plumage. The local name is that also given to sandpipers, meaning 'something which moves fast' (Keulemans 1866).]

[Tree Pipit   *Anthus trivialis*   –;?;–
                             (Linnaeus 1758)

Breeds Palearctic, winters Africa and India. Perhaps vagrant São Tomé.

One seen by P. Christy on the road between Bom Sucesso and Macambrara, Dec 1994 (Christy & Clarke 1998, P. Christy *in litt*).]

TURDIDAE

Whinchat   *Saxicola rubetra*   **VPM;–;–**
                             (Linnaeus 1758)

Palearctic migrant to tropical Africa. Vagrant Príncipe.
    One bird shot near Baía de Santo António, Nov 1865 by Keulemans (1866), who reported that it was unknown to the inhabitants.

Gulf of Guinea Thrush   *Turdus olivaceofuscus*   **RB****
                             Hartlaub 1852   **(Príncipe & São Tomé)**

Endemic to São Tomé and Príncipe, with a separate subspecies on each. Considered 'Vulnerable' by Stattersfield *et al* (1998) but subsequently only as 'Near Threatened' by BirdLife International (2000).
    The affinities of this species are not clear (see Keith & Urban 1992). Naurois (1984c) considered it close to *T. bewsheri* of Anjouan Island, Comoros, but overemphasised similarities in their habitats and ecology, and perhaps placed too much weight on the taxonomic significance of similarities in their eggs and nests (see Lack 1958) and in their barred underparts (which differ from the plain or streaked underparts of what would otherwise be seen as their closest relatives); as Naurois admits, parallel retention of a barred juvenile plumage in the adults of insular *Turdus* could be involved. Keith & Urban (1992) suggest that both species may be descended from a mainland common ancestor which is now extinct.

                             Turdus olivaceofuscus xanthorhynchus   **RB*;–;–**
                             (Salvadori 1901)

Tordo do Príncipe (Port: N 1994).

Subspecies endemic to Príncipe; one other on São Tomé.
    Discovered in 1901 at Baía do Oeste by Fea, who obtained a single adult specimen (Salvadori 1901). It has probably always been rare (see Bannerman 1914); the only other record until 1997 was by Correia in 1928 (Amadon 1953). Fea was told that it was confined to the W coast (Salvadori 1903a), but Correia collected all 4 of his examples (of which 3 are in AMNH) in the hill forests of the south (Correia 1928–29b) and stated that they lived only on the highest peaks, on the ground and low branches in dense vegetation (Correia 1928–29a). Naurois (1984c) unsuccessfully searched for the bird in the W, SW and SE; however, others had also looked for it in vain (eg Alexander: Bannerman 1914) before Correia found it. In 1988 fragmentary *T. olivaceofuscus* songs were tracked to Príncipe Drongos *Dicrurus modestus* singing quietly at Infante Dom Henrique and on Rio Papagaio (J&B). One thrush was seen and photographed on 11 June 1997, on a mesa at *c* 120 m altitude above R. São Tomé (S. d'Assis Lima *in litt*); it sang differently from São Tomé birds. Another photographed by J. Rosseel (*in litt*) on 'Morro Mesa' (A Mesa), who also reported that guides say it

is also present in the south. The 'thrush anvil' sites attributed to this species by Christy & Gascoigne (1996) are now believed to have been made by Blue-breasted Kingfishers (J. Baillie and A. Gascoigne pers comm).

Habits Correia (1928–29b) reported that it lived on the ground and understorey up to 1.5 m height in dense forest on high hills.

Turdus olivaceofuscus olivaceofuscus   –;RB*;–
Hartlaub 1852

Todo, Toldo (corruptions of Tordo), Tordo (B 1888c *et al*, Correia 1928–29a, N 1972b). Port: Tordo de São Tomé (F&V, N 1994).

Subspecies endemic to São Tomé; one other on Príncipe.

   Common and widespread in the late 19th century (Bocage 1891). At present, still common, in most types of plantation and forest throughout the island, including dry woodland in the N, and from the coast up to at least 1500 m; infrequently away from the cover of tall trees and not in open savanna; often seen on or near ground, where keeps in undergrowth, but sometimes high in trees (Snow 1950, Naurois 1984c, Günther & Feiler 1985, Jones & Tye 1988). Possibly absent from unshaded cocoa plantations (Atkinson *et al* 1994).

Habits Most often seen towards dusk by Jones & Tye (1988) and may be partly crepuscular. Wary but sometimes permits close approach (Jones & Tye 1988). Forages on the ground, searching under leaf litter (Snow 1950, Naurois 1984c); smashes snails on rocks (Atkinson *et al* 1994). Gleans underside of leaves (J&T) and eats oil-palm *Elaeis guineensis* fruit (J&B). Stomach contents include remains of snails and large insects (mainly Coleoptera), and a worm as well as fruit pulp (Snow 1950, Naurois 1984c); eats cultivated fruit and a variety of wild species, including avocado and guava, *Spondias cytherae*, *Cecropia*, *Ficus* spp. Sings much, with the quality of Blackbird *T. merula* (Snow 1950, Naurois 1984c, Günther & Feiler 1985) although the song is more monotonous (J&T). Mild alarm call a quiet 'chup'. Song and call tape-recorded (J&T, J&B, P.D. Alexander-Marrack); sonograms of both presented by Nadler (1993); calls also described by Christy & Clarke (1998). Also makes a wing-noise when flying (Snow 1950).

Breeding The nest and eggs were described by Bocage (1891, 1904) and Naurois (1972b, 1984c, 1994). Data available to Naurois (1984c) indicate a laying period from mid-Jul to Dec, with a peak Oct–Dec, and perhaps with a few pairs breeding at other times. Other specimens, collected in Jun, had poorly developed gonads (Günther & Feiler 1985), while song frequency (little in Jun, much in Jul–Aug) supports a commencement of breeding in Jul (Günther & Feiler 1985, Jones & Tye 1988). Birds seen chasing one another Jul–Sep (Snow 1950, Jones & Tye 1988).

Morphometrics of 9 trapped birds were given in Atkinson *et al* (1994).

SYLVIIDAE

Great Reed Warbler   *Acrocephalus arundinaceus*   –;VPM;–
(Linnaeus 1758)

Palearctic, wintering Africa. Vagrant São Tomé.

   Reported on São Tomé by Christy & Clarke (1998) as frequent Dec–Jan, singing in dense scrub and tall grass in coastal marshes and near the city, but the only definite record seems to be two observations of singing birds in Jan 1996 (P. Christy *in litt*).

[Chattering Cisticola   *Cisticola anonymus*   –;?;–
                                 (Müller 1855)

Afrotropical. Recorded from São Tomé, probably in error.

Hartlaub (1850, 1857) lists this species (then called *Drymoica ruficapilla*) as collected by Weiss. However, his description of the specimen fits fairly well a female or immature *Prinia molleri* and it is likely that this was the bird collected by Weiss. Perhaps Hartlaub had not seen specimens of *D. ruficapilla,* which had recently been described by Fraser (1843a). Unfortunately, both the ZMH catalogues and (almost certainly) Weiss's specimen were destroyed in World War II; the bird is no longer in the collection there (C. Hinkelmann *in litt*).]

São Tomé Prinia   *Prinia molleri*   –;RB**;–
                               Bocage 1887a

Tucli (B 1891, S 1903b); Todi, Troqui, Truqui, Truqué, Tacle-tacle (all onomatopoeic), Bate-asas (F, F&V, N 1972b, 1984b, R), Truqui-sum-Dessu, Plá-plá (M). Port: Prínia de Moller (N 1994).

Endemic to São Tomé; no close relatives apparent among *Prinia* spp of the neighbouring mainland (Naurois 1984b).

The first record of this species may have been Weiss's specimen, attributed by Hartlaub (1850, 1857) to Chattering Cisticola *Cisticola anonymus* (*qv*). *P. molleri* was subsequently described by Bocage (1887a) based on specimens collected in 1885 by Moller.

Abundant in all open habitats, including city, villages, grasslands, farms, plantations of all kinds and forest edge throughout the island, breeding at least up to 1400 m, and seen near the Pico summit at 2000 m (Snow 1950, Naurois 1984b, Jones & Tye 1988). Not a 'savanna bird' (*pace* Amadon 1953) but only frequent to common inside dense forest regrowth or undisturbed forest (Naurois 1984b, Jones & Tye 1988, J&B, Harrison & Steele 1989).

Habits Usually found low in undergrowth (Naurois 1984b, J&T) where it hunts its food mainly near or on the ground (Snow 1950, J&T); also forages higher in trees (Naurois 1984b, J&T); gleans insects under leaves and sometimes sally-gleans (J&T). Food includes Coleoptera, grasshoppers, caterpillars and vegetable matter (Naurois 1984b, J&T, Atkinson *et al* 1994). Noisy and conspicuous, often using exposed perches; found singly or in groups of up to 17 birds (when displaying) (Günther & Feiler 1985, Jones & Tye 1988, Atkinson *et al* 1994). Most observers have noted its conspicuous display (sometimes communal), calls and wingbeat sounds, which are given in flight or when perched (eg Salvadori 1903b, Bannerman 1915a, Günther & Feiler 1985, J&T); detailed descriptions were given by Snow (1950), Naurois (1984b), Nadler (1993) and Christy & Clarke (1998); Nadler (1993) includes a sonogram which shows the synchronisation of wingbeat sounds with calls. Also gives a squeaky, nasal call *yip, yip...* (J&T). The communal display is rarely observed and its function unknown; perhaps a lek (Naurois 1984b). Repetitive display-song and other calls tape-recorded (J&T, J&B, P.D. Alexander-Marrack).

Breeding Nests and eggs were described by Bocage (1904), Snow (1950) and Naurois (1984b, 1994). Copulation observed mid Jul (Jones & Tye 1988); egg-laying takes place over at least the period Aug–Feb with a peak Nov–Jan (Naurois 1984b); adults feeding young Mar–Apr (Nadler 1993); fledglings seen Feb (Harrison & Steele 1989). Three of 5 caught, Aug, were moulting (Atkinson *et al* 1994).

Morphometrics of 7 trapped birds were given in Atkinson *et al* (1994).

[Willow Warbler   *Phylloscopus trochilus*   ?;?;–
                                 (Linnaeus 1758)

Palearctic, winters Africa.

   Several individuals reported by I. Sinclair to have landed on a boat between São Tomé and Príncipe, 30 Mar 1992 (Christy & Clarke 1998, P. Christy *in litt*).]

Garden Warbler   *Sylvia borin*   –;VPM;VPM
                                (Boddaert 1783)

Palearctic, winters Africa. Vagrant São Tomé and Annobón.

São Tomé One seen above Roça Agostinho Neto Nov 2004 (C. Hjort *in litt*).

Annobon One collected Nov 1987 is now in the collection of the Estación Biológica de Doñana (J. Pérez del Val *in litt*).

Bocage's Longbill,   *Amaurocichla bocagii*   –;RB***;–
São Tomé Short-tail   Sharpe 1892a

[Vernacular unknown: N 1994]

Monotypic genus endemic to São Tomé.

   Discovered by Newton, who collected 3 males in 1890 and 1891, 2 in forest near São Miguel and one at Bindá (W) (Bocage 1904), or else one at each of these localities and the third at Juliana de Sousa (SW) (Naurois 1982). In 1928, Correia collected a pair and another female at Rio Quija, about 8 miles from Roça Jou (SW) (Correia 1928–29a,b, Collar & Stuart 1985). Two of Newton's specimens were destroyed in the 1910 fire at the Bocage Museum (Naurois 1983a); the type survives (BMNH), as do Correia's skins (AMNH).

   It was not subsequently recorded until Eccles (1988) sighted what appeared to be this species at the Rio Caué road bridge (SE) in Apr 1987. Jones & Tye (1988) twice searched the site where Eccles saw the bird, but failed to relocate it. *Contra* Naurois (1994: p. 20–21), we did not see it in 1985, never having visited the island by that date! Eccles' bird was seen in poor light (J.-J. Bolyn, pers comm) in unusual habitat and in an area where the species is not otherwise known (see Atkinson *et al* 1991) and is perhaps subject to some doubt. However, at least 2 populations were recently found in the SW and SE, on the Rios Xufexufe and Ana Chaves, with 4–6 pairs per km of river (Atkinson *et al* 1991). It was also found on the Xufexufe in 1991, by Nadler (1993), and seen or heard in early 1997 in almost every forested river basin between 400 and 600 m asl, from Formoso Grande to R. São Miguel (S. d'Assis Lima *in litt*); it occurs from there to at least the catchments of the Ió Grande and Ana Chaves (Christy & Clarke 1998). Although Atkinson *et al* (1991) did not find it along the Rio Quija, where it had last been recorded, and suggested that disturbance in the intervening period might have extinguished this population, it does still occur there (Christy & Clarke 1998). Given the species' alleged elusive nature (Correia 1928–29b, Collar & Stuart 1985, but see below), it may be even more widespread in the south. Considered 'Vulnerable, D1' (population <1000 adults) by BirdLife International (2000) on account of its apparently narrow habitat requirements. If it occurs along all streams likely to provide suitable habitat up to 500 m altitude (none has yet been seen higher than this), the total population would be 380–525 birds, whereas if it occurs up to 1000 m, the figure would increase to 680–930 (Peet & Atkinson 1994); however, it may also occur away from streams.

Habits The species appears to be mainly found in riparian habitats within primary forest, although not restricted to waterside habitats. Correia (1928–29a,b) found a pair and a female on stones and mud, and among low branches and bushes, by shallow streams in ravines in primary forest; he thought it was a rail. Atkinson *et al* (1991) noted that they were apparently territorial and foraged on gravel beds by riversides; they did not find birds more than 7–10 m away from either a river or forest stream with overhanging vegetation and moss-covered rocks. Also found in stream-beds by Nadler (1993). They also occur on fallen logs and low branches, on forested ridges up to 500 m away from waterways and up to 200 m above the valley bottoms, though always in association with temporary rivulets on the forest floor (M. Melo pers comm). The gait resembles that of a pipit *Anthus* sp (T. Gullick *in litt*, Sargeant 1994, Christy & Clarke 1998). Climbs on horizontal and inclined branches (Newton in Naurois 1982, 1994, Christy & Clarke 1998). Abraded tail feathers of some specimens have suggested to some that it might behave like a treecreeper (Naurois 1982, 1994) but this habit has not been reported from recent observations.

Foraging, other behaviour and calls were described by Atkinson *et al* (1991, 1994), Nadler (1993) and Christy & Clarke (1998); Nadler (1993) presented a sonogram. Both sexes were easily located by their loud call; territorial males sang at night; the birds were reluctant to fly, and when alarmed often ran in preference to flying.

Breeding No breeding data. Among the 3 birds found by Nadler (1993), one had grey legs, another (young?) pinkish. A group of 4 reported Aug; one captured Mar was moulting (Christy & Clarke 1998).

Morphology and systematics The affinities of this bird are not clear. It was placed *incertae sedis* among the Timaliidae by Dowsett & Forbes-Watson (1993). Sharpe (1892a) thought it resembled SE Asian babblers of the genus *Crateroscelis*. It has nine primaries and 10 rectrices, perhaps suggesting that it is an aberrant sylviid warbler (Amadon 1953), and possibly closest to the *Macrosphenus* longbills (but see Naurois 1982). The lack of a tenth primary suggested to Naurois (1982) that it might be descended from a northern migrant sylviid, or even, considering this and the protruding rachis of the tail feathers, that it could be a relic of the ancient stock which gave rise to the Neotropical dendrocolaptid-furnariid assemblage! The protruding rachis in *Amaurocichla* is narrow and appears to be the result of abrasion (AT at BMNH), as in the Furnariidae (whereas it is thickened and protrudes as a developmental feature in the Dendrocolaptidae). More reasonably, Naurois (1994) drew attention to the narrowing of the mid-part of the beak, similar to that of *Illadopsis*.

Morphometrics of 2 trapped birds given by Atkinson *et al* (1994).

MUSCICAPIDAE

Spotted Flycatcher   *Muscicapa striata*   **VPM;VPM;VPM**
                               (Pallas 1764)

Port: Papa-moscas-europeu (F&V). Sp: Papamoscas gris (Bas).

Palearctic migrant to Africa. Perhaps regular all three islands.

Príncipe Two adult males (probably *M. s. striata* from Europe) collected at Roça Infante Dom Henrique (SE) and Baía de Santo António (NE), Nov 1954 (Frade 1958, 1959). In Mar 1992, one reportedly landed on a boat between Príncipe and São Tomé (Christy & Clarke 1998).

São Tomé A sighting reported by Reinius (1985).

Annobón A male, identified by Amadon as the race *M. s. balearica*, collected in Nov 1955 (Basilio 1957). The stomach contained mainly ants, with beetles and small locusts (Basilio 1957).

MONARCHIDAE

São Tomé Paradise Flycatcher   *Terpsiphone atrochalybeia*   –;RB**;–
(Thomson 1842)

Tomé-gagá, Zé-zé, Jíji (S 1903b, B 1904, F&V, N 1994, C&C). Port: Papa-moscas de São Tomé (N 1994).

Endemic to São Tomé; forms a superspecies with *T. viridis* of mainland Africa (detailed discussion in Naurois 1984a). The correct spelling of the specific name is as given here, not *atrochalybea* (see Thomson 1842). The '*Muscipeta*' *melampyra* reported from São Tomé by Bocage (1867) is the female of *T. atrochalybeia*.

   Common to abundant in plantations, forest regrowth and forest; often seen at forest edge, clearings, rivers, etc, but perhaps because more conspicuous there (Jones & Tye 1988). Uncommon in northern savannas, where found in patches of closed-canopy dry woodland, maize fields and grassy places (Naurois 1984a, Jones & Tye 1988, Atkinson *et al* 1994). Occurs up to about 1600 m, eg at Lagoa Amélia (Bannerman 1915a, Snow 1950, Naurois 1984a). Reportedly suffered a decline in plantations (only) due to heavy pesticide use therein in the 5 years prior to independence (Naurois 1984a) but has subsequently recovered to pre-1971 levels (Jones & Tye 1988) and reaches 125 birds per km² in cocoa with shade trees (Atkinson *et al* 1991); absent from plantations without shade trees (Atkinson *et al* 1994).

Habits Usually seen low in the understorey in twos (2 males, 2 females, male and female), threes (male plus 2 females) or occasionally more birds together (Snow 1950, Naurois 1984a, Günther & Feiler 1985, Jones & Tye 1988, J&T). Often tame and curious, approaching observer closely (Jones & Tye 1988). Vocalisations were described by Snow (1950), Naurois (1984a) and Christy & Clarke (1998); chipping and buzzing calls tape-recorded (J&T, J&B, P.D. Alexander-Marrack). Stomach contents include Coleoptera and (probably) Hemiptera (Naurois 1984a); food caught by sally-gleaning and aerial sallying (J&T, Naurois 1994). One adult male seen by J&B diving repeatedly onto the surface of a still pool in the Ribeira Afonso, possibly drinking but more likely picking up surface insects. One followed a foraging São Tomé Weaver *Ploceus sanctithomae*, apparently catching insects disturbed by the weaver (Eccles 1988).

Breeding Nests and eggs were described by Bocage (1891, 1904), Bannerman (1915a), Amadon (1953) and Naurois (1972b, 1984a, 1994); photograph of nest in Naurois (1984a). Laying period Jul–Jan, with a peak Oct–Dec (Naurois 1984a). Nest-building Aug (P.D. Alexander-Marrack *in litt*); nest with 2 nestlings Nov (M. Melo pers comm); dependent fledglings Feb and Aug (Harrison & Steele 1989, P.D. Alexander-Marrack *in litt*).

Morphometrics of 7 trapped birds were given in Atkinson *et al* (1994).

Annobón Paradise Flycatcher   *Terpsiphone smithii*   –;–;RB**
                                    (Fraser 1843b)

Bibí (B 1893a, Bas). Sp: Moscareta de Smith (Bas).

Endemic to Annobón.

Often regarded (eg by Dowsett & Forbes-Watson 1993, Urban *et al* 1997) as a subspecies of *T. rufiventer*, which has races on Bioko and mainland, and with which it forms a superspecies.

Common and widely distributed (Bocage 1893a, 1903, Schultze 1913, Basilio 1957, Fry 1961, Harrison 1990, Pérez del Val 2001), in forest, woodland and cultivated areas, not away from 'bush' (Fry 1961, Harrison 1990). Perhaps most common around the crater lake (Bocage 1893b, Salvadori 1903c) and in woodland along streams (Basilio 1957). Considered 'Vulnerable, D1, D2' (population <1000 adults, range <100 km² or <5 locations) by BirdLife International (2000).

Habits Often in pairs, occasionally in groups of up to 5 (Basilio 1957). Calls and song (given by both sexes) described by Fry (1961) and Harrison (1990); Fry reported it not singing much in Jul and Aug. Said by local people to reveal their secrets to the missionaries (Barrena 1911). Sallies after aerial insects but stomach contents also include seeds (Basilio 1957, Fry 1961).

Breeding Two Sep males had regressed testes (Basilio 1957); breeding unknown.

TIMALIIDAE
(incertae sedis)

Dohrn's Thrush-babbler   *Horizorhinus dohrni*   RB***;–;–
                               (Hartlaub 1866)

Sibi-fixe (Dohrn 1866, K); Rouxinol (N 1994); Tchibi-fixa (C&C); Chibi-peito-branco (M). Port: Rouxinol do Príncipe (F, C&C).

Monotypic genus endemic to Príncipe.

Abundant in cocoa plantations and forest regrowth (Jones & Tye 1988) and in other bushy country; perhaps less common in coconut plantations (Snow 1950). Also occurs in villages, and at all altitudes (Keulemans 1866). Has apparently always been abundant (eg Keulemans 1866, Bocage 1903, Snow 1950).

Habits Mainly in the understorey and bases of the crowns of taller trees, but also forages low among lianas, like a bulbul (Pycnonotidae) or forest robin *Cossypha* (Jones & Tye 1988). Sometimes in parties of 2–8 birds, occasionally more than 20 (Keulemans 1866). Normally secretive but sometimes very tame; noisy, frequently singing and calling (Keulemans 1866, Snow 1950, Jones & Tye 1988); calls and loud, beautiful song described by Keulemans (1866), Snow (1950) and Christy & Clarke (1998); also a persistent mewing call and a 'ticking' contact call like that of a *Sylvia* warbler (J&T). Mewing, trill and song tape-recorded (J&T, J&B). Aggressive, pursuing conspecifics and other small birds (Keulemans 1866), but occasionally in mixed parties with Príncipe Golden Weaver *Ploceus princeps* and Príncipe Glossy Starling *Lamprotornis ornatus* (J&T). Pokes into bark crevices (J&T). Food mainly caterpillars, ants, beetles, other insects and small snails, but also berries and seeds (Keulemans 1866, Snow 1950, Snow *in* Bannerman 1930–51 vol. 8).

Breeding Nest and eggs described by Dohrn (1866), Keulemans (1866), Snow (1950) and Naurois (1972b, 1994). Snow found them breeding Jun–Sep; Naurois (1994) thought that they bred Jun–Jul and Sep–Jan (May–Jun and Sep–Feb according to Naurois 1972b); probably double-brooded (Keulemans 1866). Young fed mainly on berries (Keulemans 1866).

Morphology and systematics Most individuals seen by Jones & Tye (1988) had a strong yellow wash on the underparts, which is not evident in museum specimens, nor in most illustrations and descriptions (eg see Bannerman 1930–51 vol. 4) and which perhaps fades after death (*cf Lanius newtoni*). It is shown in the plate in Dohrn (1866) and seems to have been noted by Correia (1928–29b), for he termed the bird 'yellow balle' (yellow belly). Naurois (1994) thought that the yellow wash appeared during the breeding season only. Morphometrics of 3 trapped birds given by Atkinson *et al* (1994).

The affinities of this species are unclear; it was placed *incertae sedis* among the Timaliidae by Dowsett & Forbes-Watson (1993) and Naurois (1994), mainly because of the narrowing of the middle of the beak, similar to that of *Illadopsis*, but in other respects it seems closer to the Sylviidae or Muscicapidae. There are similarly hard-to-place muscicapids (*sensu lato*) on the mainland; all may be relics of an old, undifferentiated muscicapid stock (Moreau 1966).

NECTARINIIDAE

Príncipe Sunbird,　　　　*Anabathmis hartlaubii*　　　　**RB**;–;–**
Hartlaub's Sunbird　　　　J. Verreaux 1857

Sibi-barbeiro, Sibi-boca-longa (K); Xibi-sangue (B 1903); Beija-flor (Dohrn 1866, K, F&V). Port: Beija-flor do Príncipe (C&C).

Endemic to Príncipe.

Considered fairly common except in dense hill forest by Dohrn (1866), Keulemans (1866) and Snow (1950). At present common to abundant in plantations (including coconut), forest regrowth (Snow 1950, Jones & Tye 1988), farmland and gardens (Keulemans 1866, Atkinson *et al* 1991) and the forests of the lowland SW (Christy & Clarke 1998). Occurs to at least 150 m (J&T).

Habits Usually seen in pairs or small groups: probably family parties (Keulemans 1866, Snow 1950, Jones & Tye 1988). Calls and song were described by Keulemans (1866), Snow (1950) and Christy & Clarke (1998); tape-recorded (J&T, J&B). Most vocal in late afternoon (Jones & Tye 1988). Food includes small insects, particularly ants and aphids, also nectar (including that of banana) and fruit pulp (Keulemans 1866, Snow 1950, Naurois 1994). Foraging behaviour described by Keulemans (1866) and Snow (1950). Pokes in ends of broken twigs (J&T).

Breeding Nests were described by Dohrn (1866), nests, egg and immatures by Keulemans (1866, and *in* Shelley 1896–1912), nests and eggs by Naurois (1972b, 1994). Immatures said to have been collected Mar (Bannerman 1914) were actually Olive Sunbird *Cyanomitra olivacea* (Bannerman 1930–51 vol. 6). Dohrn (1866), from 6 months' observations, thought the breeding season must be long, and was told by locals that the birds 'keep their nests during the whole year'. Keulemans (1866) stated that pairs feeding young could be seen at all times of year and thought there was 'no fixed breeding time', but with a probable peak about Aug (see Keulemans *in* Shelley 1896–1912, where further breeding information is given). A nesting pair was found Feb by Harrison & Steele (1989). Naurois (1994) gave breeding season as Nov–Feb, perhaps Sep–Mar (Oct–Feb in Naurois 1972b), but without presenting data.

Morphometrics of a trapped bird given by Atkinson *et al* (1994).

São Tomé Sunbird, *Anabathmis newtonii* –;RB**;–
Newton's Sunbird Bocage 1887a

Xêle-xêle (B 1888c, 1904); Cerelé, Beija-flor (N 1972b); Selêlê (R); Chtrêlê (M). Port: Beija-flor de peito amarelo (F&V, C&C); Beija-flor-pequeno de São Tomé (N 1994).

Endemic to São Tomé.

Common in all habitats with tall trees, including city, plantations and secondary forest; perhaps less common in the SW; abundant in high-altitude coffee plantations of Monte Café area, in overgrown plantations between Lembá and Bindá, W coast, and in forest up to the summit of the Pico; not found in northern savannas (Correia 1928–29b, Snow 1950, Jones & Tye 1988, J&B, Harrison & Steele 1989) but occurs in coffee and cocoa without shade trees (Jones & Tye 1988, Atkinson *et al* 1991).

Habits Song and calls were described by Snow (1950) and Christy & Clarke (1998); additional calls *bink* like Chaffinch *Fringilla coelebs*, a rising *zee-eep* and a squeaky, grating *zhreeeuw* (J&T). Calls and song tape-recorded (J&T, P.D. Alexander-Marrack); a sonogram of one call-type presented by Nadler (1993). Singing frequently in Jul and Aug (J&T), when both sexes responded markedly to playback (J&B). In breeding display, beats wings while singing furiously from perch (Naurois 1994). Often found in parties and accompanying other species, including the São Tomé Speirops *Speirops lugubris* and Príncipe White-eye *Zosterops ficedulinus* (Jones & Tye 1988). Playback of alarm calls attracted 5 of the latter species (J&B). Gleans insects from under leaves (J&T, Nadler 1993). Congregates in groups of up to 30 at flowering *Erythrina* trees, Jul–Aug (Atkinson *et al* 1991).

Breeding Nest and eggs described by Bocage (1891, 1904), Amadon (1953) and Naurois (1994). Nest-building Dec–Jan (Christy & Clarke 1998); 3 or 4 nests with eggs, Oct–Feb (Correia 1928–29b, Amadon 1953); a juvenile with yellow gape, Apr (Eccles 1988); dependent young Jan (Christy & Clarke 1998); pair with young, Sep (Frade 1958); uniform grey individuals, thought to be juveniles, Jun (Günther & Feiler 1985); family parties, Jul–Aug (J&T). Laying period mid-Aug to Jan, according to Naurois (1994), although no data given; breeding period Oct to early Mar according to Naurois (1972b). Naurois (1994) also stated that it is sometimes parasitised by São Tomé Weaver *Ploceus sanctithomae* but we know of no evidence for this extraordinary suggestion (error for Emerald Cuckoo *Chrysococcyx cupreus?*).

Morphometrics of 5 trapped birds given by Atkinson *et al* (1994).

Giant Sunbird *Dreptes thomensis* –;RB***;–
  Bocage 1889b

Zom-zom (B 1904); Cerelé-de-obó (N 1972b); Selêlê-mangotchi (R). Port: Beija-flor-gigante (F&V, C&C); Beija-flor-preto de São Tomé (N 1994).

Endemic to São Tomé.

Was apparently widespread towards the end of the 19th century, when Newton collected many 'dans les forêts de S. Miguel' (SW) (Bocage 1904), presumably forests above the roça. Newton also collected some at 'Mussa Cadi' (presumably Mussucavú) and 'Budo-Tap-ana' (see Dwarf Ibis *Bostrychia bocagei* account): if the latter is Ubabudo, then the species once also inhabited forests at moderate altitude in the E. More recently found at low altitude in the S and SW, where locally common (Correia 1928–29a, Naurois 1983a, Atkinson *et al* 1991, Sargeant 1994) and at high altitudes, especially around Lagoa Amélia and up to the Pico, where frequent to common (Correia 1928–29b, Snow 1950, J&B, Atkinson *et al* 1991, Nadler 1993, Christy & Clarke 1998); only above about 800 m in the east (Naurois & Castro Antunes 1973). Mostly in primary forest but found in cultivated areas near Bom Sucesso (1100 m) in 1988 (J&B). Considered 'Vulnerable, D1' (population <1000 adults) by BirdLife International (2000).

Habits Mostly in the canopy or subcanopy but sometimes descends lower (Christy & Clarke 1998). Continually on the move from tree to tree (J&B). Behaves like a treecreeper *Certhia* or woodhoopoe Phoeniculidae, poking into bark (Correia 1928–29b, Christy & Clarke 1998); also gleans from leaves (Correia 1928–29b) and takes nectar of bananas and other flowers (Snow 1950, Sargeant 1994) when may hover by flowers (Correia 1928–29b). Has a loud 3–5-note call, a harsh chipping call (Snow 1950, J&B) and a rambling, dusk song (Atkinson *et al* 1994); further descriptions in Christy & Clarke (1998); calls tape-recorded (J&B); apparently territorial, responding strongly to playback (PJJ).

Breeding Correia found 3 nests, with eggs laid Dec and Jan, one at Rio Quija and the others at Roça Jou: they were described by Correia (1928–29b) and Amadon (1953). Naurois (1994) also described nest and eggs, and gave breeding season as Sep–Jan (Oct–Jan in Naurois 1972b), but it is not clear whether these were based on his own observations. Perhaps polygynous (Atkinson *et al* 1991).

Morphometrics of 5 trapped birds given by Atkinson *et al* (1994).

Olive Sunbird	*Cyanomitra olivacea*	RB;–;–
	(A. Smith 1840)	

Beija-flor (Dohrn 1866, N 1972b). Port: Beija-flor-oliváceo (F&V, C&C). Local names said by Keulemans (1866) and Naurois (1972b) to be the same as for Príncipe (Hartlaub's) Sunbird *Anabathmis hartlaubii* (*qv*), but later termed Siwie-barbeiro-grande by Keulemans (*in* Shelley 1896–1912).

Afrotropical. Resident Príncipe, where birds belong to mainland *C. o. cephaelis*, not (as formerly thought) to the Bioko endemic *C. o. obscura* (Tye & Macaulay 1993).

Considered common in dense forests of uninhabited parts by Keulemans (1866) but strangely not obtained by Correia (Amadon 1953). Now common in plantations and forest regrowth, usually in groups of 2–5 birds (Jones & Tye 1988). Perhaps lives (or lived) at higher altitudes than reached by the Príncipe Sunbird (Dohrn 1866); certainly not restricted (*pace* Naurois 1983a) to 'alt. basse'.

Habits Said to live high in trees (Keulemans *in* Shelley 1896–1912) but also occurs in low shrubs (J&T) and low in overgrown cocoa plantations (J&B). Sometimes alongside Príncipe Sunbird (Naurois 1983a, J&T). Eats invertebrates (termites, spiders, aphids), banana pulp, nectar (Naurois 1994). Song described by Keulemans (1866, *in* Shelley 1896–1912) and Christy & Clarke (1998); song (tape-recorded) rises and falls, speeds up and slows like a squeaky wheel cranked at different speeds *pi, pi, pi, pi, pi-pi-pi-pi, pi* (J&T, J&B).

Breeding Nests and eggs described by Naurois (1994). Immatures collected Mar (listed as Príncipe Sunbird *Anabathmis hartlaubii* by Bannerman 1914, see Bannerman 1930–51 vol. 6 p. 226 footnote). Keulemans (1866) thought them paired the whole year. Naurois (1994) gave breeding season as Nov–Jan (perhaps to Apr), although earlier (1972b) he stated only that eggs were found in Feb.

ZOSTEROPIDAE

Príncipe White-eye   *Zosterops ficedulinus* **RB** (Príncipe & S. Tomé)**
                                    Hartlaub 1866

Tchili-tchili (R). Port: Olho-grosso-pequeno de São Tomé e Príncipe (N 1994); Olho-branco de São Tomé (C&C).

Endemic to São Tomé and Príncipe, with a separate subspecies on each. Classified as 'Vulnerable, D1' (population <1000 adults) by BirdLife International (2000).

                Zosterops ficedulinus ficedulinus   **RB*;–;–**
                Hartlaub 1866

Owee-gapao (K); Xibi-Tete (B 1903).

Subspecies endemic to Príncipe; one other on São Tomé.

   Considered fairly common in the mid-19th century in forests of the uninhabited hilly interior, less so at lower altitudes (Dohrn 1866, Keulemans 1866, Bannerman 1914), but few records since. One was collected by Jacintho António de Sousa in 1880 (José Augusto de Sousa 1888); Newton found it in 1887 at 'Sindy' (Bocage 1903), probably Sundi, NW (Collar & Stuart 1985); Fea encountered it at Bahia do Oeste (Salvadori 1903a), probably the large bay below Roça Oeste (W). Correia collected only 2 (Correia 1928–29a, Amadon 1953), while Frade (1958) and Snow (1950) failed to find it. Naurois (1983a) saw only one, at A Mesa (500 m, SW) in 1972 or 1973 (see Naurois 1972b), and considered that it may have suffered a decline in the 1960s and 1970s. However, it seems to have been rare even before this. Still survives in the forested highlands and S and SW (Christy & Clarke 1998), and is perhaps still common there, though the population must be small and in danger of chance extinction (Jones & Tye 1988). Considered 'Endangered' by Gascoigne (1995). At least formerly occurred in plantations as well as forest (Keulemans 1866, Correia 1928–29b), like the race on São Tomé.

Habits Call said to be very similar to that of Príncipe Speirops *Speirops leucophaeus* (Keulemans 1866). Forms flocks, which sometimes join flocks of Príncipe Speirops (Keulemans 1866). Eats berries and small insects (Keulemans 1866).

Breeding Keulemans (1866) recorded breeding in Sep 'and later', and described the nest and eggs.

                Zosterops ficedulinus feae   **–;RB*;–**
                Salvadori 1901

Selele (B 1891); Pastelim, Dã-buto (B 1904); Tchili-tchili (R); Tchelé-tchelé (C&C); Neto-d'olho-grosso (M).

Subspecies endemic to São Tomé; one other on Príncipe.

   At least formerly, occurred throughout the island, although most early records were from low altitudes (Collar & Stuart 1985). Alexander's locality 'Zalma' is probably As Almas, which is in the lowlands S of the city (Bannerman 1915a, Correia 1928–29a,b, Naurois 1983b) and is not the highland Zampalma (*contra* Collar & Stuart 1985). Now apparently most numerous at middle altitudes (Naurois 1983a, Jones & Tye 1988), though occurs lower on the W coast (*cf* Maroon Pigeon *Columba thomensis*) (Jones & Tye 1988). Found to be common by Correia in the 1920s (Amadon 1953) but seriously declined in the 1960s and early 1970s (Naurois 1983a), when very rare below 800 m and in plantations. In 1987 it appeared to have recovered somewhat, and was common to abundant in small groups in forest regrowth and cocoa plantations near Monte Café, Nova Moca and Bombaim and in forest regrowth on W coast north of Bindá, but uncommon in cocoa plantations with tall trees near Ribeira Peixe, in the lowland SE (Jones & Tye 1988). Also found recently at Lagoa Amélia, Santa Catarina, Rio Ana Chaves and Cascata (Alexander-Marrack 1990, Sargeant & Alexander-Marrack 1990, Atkinson *et al* 1991, Nad(Newton's) ler 1993, Sargeant 1994), and considered common at Bom Sucesso and between Bombaim and Angolares in Sep–Oct 1997 (P. Kaestner *per* P. Atkinson *in litt*). However, still considered 'very rare' by Atkinson *et al* (1994), and 'rare' by Christy & Clarke (1998). Recorded in northern dry woodland by Harrison & Steele (1989). Still local, despite some evidence of recovery.

Habits Sometimes forms mixed parties with one or more of the following: São Tomé Speirops *Speirops lugubris*, São Tomé Sunbird *Anabathmis newtonii*, São Tomé Prinia *Prinia molleri*, São Tomé Paradise Flycatcher *Terpsiphone atrochalybeia* (Naurois 1972b, Günther & Feiler 1985, Jones & Tye 1988). Fea noted that this subspecies has 'the same habits' as the São Tomé Speirops (Salvadori 1901); these 2 species forage in the same way in mixed parties, gleaning insects from the underside of leaves (J&T).

Call a quiet chipping, similar to that of the São Tomé Speirops but perhaps quieter and higher-pitched; also a brief silvery trill, typical of other *Zosterops* spp but shorter and squeakier than *Z. senegalensis* (AT); further descriptions in Christy & Clarke (1998). *Pace* Christy & Clarke (1998), calls have been tape-recorded (J&B, P.D. Alexander-Marrack).

Breeding Fledglings seen Feb (Harrison & Steele 1989); breeding otherwise unknown.

Annobón White-eye   *Zosterops griseovirescens*   –;–;RB**
                               Bocage 1893a

Bichili (B 1893a, S 1903c); Bichil (Bas; Barrena's 'Bidúl' is probably a misprint for this); Zostérope de Annobon (Sp: Bas); Olho-grosso de Ano-Bom (Port: N 1994).

Endemic to Annobón.
   Abundant everywhere with cover, including all types of forest, wooded and bushy habitats, farmland and villages, from sea-level up to the crater lake (Bocage 1893a, 1903, Salvadori 1903c, Schultze 1913, Bannerman 1915b, Basilio 1957, Fry 1961, Harrison 1990, Pérez del Val 2001). Classified as 'Vulnerable, D2' on account of its small range (<100 km²) by BirdLife International (2000).

Habits Rendered conspicuous by its calls and song (Bocage 1903, Fry 1961), which were described by Fry (1961). Occurs in pairs or small groups (Basilio 1957, Harrison 1990). Gleans vegetation at all heights for insects (Harrison 1990). Stomach contents included more insects than fruits, and especially ants (Basilio 1957). One trapped bird weighed 11 g (Harrison 1990).

Breeding Nests with eggs and young, Nov and Dec (Bannerman 1915b, Basilio 1957); fledglings seen commonly in Feb (Harrison 1990). Nests and nest-sites were described by Basilio (1957) and Fry (1961), and eggs, young and behaviour of adults at the nest by Basilio (1957).

| **São Tomé Speirops** | *Speirops lugubris* | –;RB**;– |
| | (Hartlaub 1848) | |

Ué-glosso (S 1903b); Mandinha (B 1904); Olho-branco (N 1972b); Olho-grosso (N 1972b, R). Port: Olho-branco-sombrio (F&V).

Endemic to São Tomé: *pace* Dowsett & Forbes-Watson (1993) and Eck (1995), and in accordance with Wolters (1983) and Fry *et al* (2000), we treat the form found in Cameroon as a separate species, *S. melanocephalus*. The two differ in structural characters, with the much larger *lugubris* possessing a longer tail and more pointed wing (Feiler & Nadler 1992) and more closely resembling *S. leucophaeus* in these respects (Eck 1995). These three species may form a superspecies, with *melanocephalus* differing in structure, and *leucophaeus* in coloration, from *lugubris*. *S. brunneus* of Bioko differs from this group in both structure (long beak and tail) and dark coloration, and perhaps represents an earlier isolation event (Eck 1995).

Common to abundant, usually in groups of up to *c* 25, in tall trees and understorey of plantations, forest regrowth and forest; in N found in plantations and patches of woodland; from sea-level up to the summit of the Pico (Salvadori 1903b, Snow 1950, Günther & Feiler 1985, Eccles 1988, Jones & Tye 1988). Formerly on Ilhéu das Rolas (Bocage 1891, 1904) but now apparently extinct there (Atkinson *et al* 1991).

Habits Members of parties continually give quiet, trilling contact calls, and the distinctive, loud song (Snow 1950, Christy & Clarke 1998) is often heard (Salvadori 1903b, Bannerman 1915a, Jones & Tye 1988); sometimes imitates other birds (Atkinson *et al* 1991). Calls and song tape-recorded (J&T, J&B, P.D. Alexander-Marrack); sonogram of contact call presented by Nadler (1993). Forms parties with other species, including Giant Weaver *Ploceus grandis*, São Tomé Paradise Flycatcher *Terpsiphone atrochalybeia*, São Tomé Oriole *Oriolus crassirostris* and Príncipe White-eye *Zosterops ficedulinus* (Jones & Tye 1988, Atkinson *et al* 1991). Distress call of captured bird attracted conspecifics as well as Príncipe White-eye, Gulf of Guinea Thrush *Turdus olivaceofuscus*, São Tomé Prinia *Prinia molleri*, São Tomé (Newton's) Sunbird *Anabathmis newtonii* and São Tomé Weaver *Ploceus sanctithomae* (J&B). Gleans leaves (Eccles 1988, J&T) and catches insects, including a caterpillar (Atkinson *et al* 1991) but stomach contents of one bird were of vegetable origin (Snow 1950) and recorded eating small berries (Correia 1928–29b, P.D. Alexander-Marrack *in litt*, Atkinson *et al* 1994).

Breeding Nest and egg described by Bocage (1891, 1904) and Naurois (1994). Data presented by Günther & Feiler (1985) and Feiler & Nadler (1992) suggest a peak moulting season about Jun, and a main breeding season ending about Jun, although birds with enlarged gonads were found Feb–Mar, Jun–Jul and Sep. Birds thought to be in smaller groups in Feb and Apr than at other times, suggesting breeding then (Eccles 1988, Harrison & Steele 1989). One carrying nest material, Dec (Sargeant & Alexander-Marrack 1990); nest-building Dec–Jan (Christy & Clarke 1998). In contrast to all these observations, Naurois (1994) gave breeding season as Jul–Dec, although without presenting supporting data, whereas Naurois (1972b) quoted it as Oct–Feb, perhaps to Apr; the breeding season may in fact be rather ill-defined.

Mean body weight of 14 birds, collected Jun, was 17 g (Günther & Feiler 1985). Morphometrics of 11 trapped birds were given in Atkinson *et al* (1994).

| **Príncipe Speirops** | *Speirops leucophaeus* | RB**;–;– |
| | (Hartlaub 1857) | |

Sibi-de-sorli, Sorli (K, R) (*sorli* is cassava); Tchiliquito (J&T, C&C); Peito-branco (C&C). Port: Olho-branco do Príncipe (N 1994).

Endemic to Príncipe. May form a superspecies with *S. melanocephalus* of Mt Cameroon and *S. lugubris* (*qv*).

Keulemans (1866) considered it very common throughout the island; Newton collected many (Bocage 1903), as did Correia in the 1920s (Amadon 1953). However, it was considered local by Dohrn (1866) and Snow (1950). Naurois (1972b) considered it 'peu abondant... 250 m et au-dessus' in the 1970s, whereas later (1983a) he described it (writing about the same period) as 'abondant; toutes altitudes'. In 1987 and 1988, frequent in parties of up to a dozen birds (up to 40 according to Naurois 1994) in plantations, perhaps commoner in the S (Jones & Tye 1988, J&B). Occurs on Ilhéu Bom-bom (Atkinson *et al* 1994). These notes might indicate that the population undergoes fluctuations, and that it may have declined in the early 1970s. Occasionally killed for food (Keulemans 1866, Jones & Tye 1988). Was not considered threatened by Collar & Stuart (1985) but, considering its small population and vulnerability to plantation development and pesticide use, Jones & Tye (1988) recommended its classification as threatened. Collar *et al* (1994) subsequently classified it 'Vulnerable, C2b' (single declining population of <10,000 adults) but it is again currently listed as 'Near Threatened' (BirdLife International 2000).

Habits Calls and foraging behaviour were described by Keulemans (1866), Snow (1950) and Christy & Clarke (1998). Song a repetitive jumble of chipping notes, lasting *c* 3 sec; this and trilling call tape-recorded (J&T, J&B); further description in Christy & Clarke (1998). Food includes berries, seeds and other vegetable matter, spiders and insects (Keulemans 1866, Snow 1950).

Breeding Nest and eggs described by Dohrn (1866) and Keulemans (1866). Birds with enlarged gonads collected Feb and Jul; birds in moult Apr, Jun–Jul (Feiler & Nadler 1992). Pair building nest, Jan (Christy & Clarke 1998). Breeding season may include June–Sep (Dohrn 1866, Keulemans 1866, Snow 1950), or Sep–Nov (Naurois 1972b); probably double-brooded (Keulemans 1866). Young fed mainly on blue berries (Keulemans 1866). Keulemans included some notes which suggest co-operative breeding.

LANIIDAE

São Tomé Fiscal,　　　　　*Lanius newtoni*　　　　　　　　　　–;RB**;–
Newton's Fiscal　　　　　　Bocage 1891

Zana (B 1904). Port: Picanço de São Tomé (N 1994).

Endemic to São Tomé. Forms a superspecies with *L. collaris*.

Discovered by Newton, who collected one at São Miguel, 3 at Rio Quija (SW), one at '*Zungui* dans l'intérieure de *Iogo-Iogo*' (identified as extreme S by Collar & Stuart 1985) and one at an unspecified locality (Bocage 1904). Bocage writes of the last specimen, '... un jeune, sans indication de sexe ni de localité. Nom vulg. *Zana* dans les *Angolares* (Newton)'. This local name for the bird (used by the Angolares people of the SE coast) was attached to the preceding sentence by Bannerman (1930–51 vol. 5), who assumed it to be the name of a locality. This produced an apparent inconsistency in what Bocage wrote, which was noticed but not solved by Collar & Stuart (1985). Thus, the 'untraceable' locality 'Zana in Angolares, on the SE coast' referred to by Collar & Stuart (1985) does not exist. Correia collected 13 birds in 1928, 12 of them in primary forest at Roça Jou and Rio Quija in the SW and one in the hills above Rio Ió Grande (Correia 1928–29a,b, Collar & Stuart 1985). Not known to Portuguese residents to whom Correia showed his specimens (Correia 1928–29b), nor (more recently) to knowledgeable hunters at Santo António and Santa Catarina (Atkinson *et al* 1991, PJJ) and was sometimes considered extinct (Collar & Stuart 1985). However, in July 1990 one was trapped in primary forest at 300 m along a tributary near the source of the Rio Xufexufe (Atkinson *et al* 1991) and in 1991 one was seen at 220 m on a ridge-top plateau above the Rio Xufexufe (T. Gullick *in litt*, Sargeant 1994). A single bird was seen by P. Christy in Jan 1996 at *c* 600 m in primary forest south of the mountain Formoso Pequeno (Bombaim area), near where Correia first saw the species in 1928 (GGCG 1996a). In 1997, a pair was watched at the edge of a clearing, at a height of 5 m to canopy level, at *c* 200 m altitude near the R. Ió Grande, and another bird at 100 m altitude near R. Martim Mendes (S. d'Assis Lima *in litt*). In 1999, 5 individuals were found farther to the southeast at three localities, at 180 m and 210 m altitude just west of the R. Ió Grande near the R. Miranda Guedes and R. João (Schollaert & Willem 2001). Now considered to be frequent within its restricted range (M. Melo pers comm). Listed as 'Critically Endangered, D1' (population <50 adults) by BirdLife International (2000) despite the current lack of information on its population size and distribution.

Habits Mistakenly called a savanna bird by Chapin (*in* Amadon 1953), an error perpetuated by Günther & Feiler (1985), this is a true forest-dwelling fiscal. The recent records and all of Correia's specimens were in dense or virgin forest, 3 of them at 1060 m altitude (Collar & Stuart 1985 *contra* Naurois 1988b). It appears to be a mid-to low-storey species, often skulking in low bushes but sometimes on more open vantage points; one foraged on boulders in a stream (Atkinson *et al* 1991); another caught small beetles on the ground (Sargeant 1994); foraging behaviour described by Christy & Clarke (1998). Apparently rather silent, but one bird emitted a low squawk on several occasions and when handled made a scolding churring sound (Atkinson *et al* 1994); song and calls described by Christy & Clarke (1998) and tape-recorded by Schollaert & Willem (2001).

Breeding All those collected by Correia (May, Nov and Dec) had small gonads except 2 males collected Nov–Dec; 2 females collected Nov–Dec had traces of juvenile plumage (Collar & Stuart 1985).

Morphology Correia (1928–29b) called the bird 'Yellow-balle' (yellow belly); in life it has a lemon wash on the underparts (Atkinson *et al* 1994), which apparently fades to white after death (BMNH), as in *Horizorhinus dohrni* (*qv*). Morphometrics of a trapped female given by Atkinson *et al* (1994).

Lesser Grey Shrike	*Lanius minor* Gmelin 1788	**VPM;–;VPM**

Port: Picanço-cinzento (F&V).

Palearctic migrant to Africa. Vagrant, Príncipe, Annobón.

Príncipe One collected at Roça Esperança (now renamed Porto Real), Nov 1954 (Frade 1958). Keulemans (1866) reported that shrikes, which he tentatively attributed to *L. excubitor*, were present in the higher, forested regions, only Nov–Jan. Salvadori (1903a) suggested that Keulemans' birds were probably *L. minor*. Keulemans reported that they were known to, but not named by, many of the inhabitants.

Annobón A shrike seen in August 2000 and reported as a Southern Grey Shrike *Lanius meridionalis* (Pérez del Val 2001) was in fact *L. minor* (J. Pérez del Val *in litt*); 8 similar birds were seen in the same place by local inhabitants in 1999 (J. Pérez del Val *in litt.*).

Red-backed Shrike *Lanius collurio* –;VPM;VPM
  Linnaeus 1758

Palearctic migrant to Africa. Vagrant, São Tomé and Annobón.

São Tomé The only record is of a juvenile observed in the city, Nov 1988 (S. Reinius *in litt*).

Annobón One juvenile trapped by J. Juste in 1998 (J. Pérez del Val *in litt*).

ORIOLIDAE

São Tomé Oriole *Oriolus crassirostris* –;RB**;–
  Hartlaub 1857

Papafigo (Vieira 1887, B 1888c and other refs, N 1972b, R, C&C); Carniceiro (Vieira 1887); Oriola (N&CA); Joãobobo (C&C). Port: Papa-figos de São Tomé (F&V, N 1994).

Endemic to São Tomé. Forms a superspecies with *O. brachyrhynchus* of W Africa (Amadon 1953, Naurois 1984d).

   Common in forest edge and regrowth between Lembá and Bindá, W Coast, with 1–2 pairs per 25 ha (Naurois 1984d, Jones & Tye 1988) and at Cascata (Nadler 1993); frequent in similar habitat near Ermelinda, SE (Jones & Tye 1988). Correia (1928–29b) met with it (his 'yellow obó thrush') at low altitudes near Rio Ió Grande as well as higher up, but considered it restricted to virgin forest and not common. Found from near sea-level up to *c* 1600 m in primary forest in the SW and at Rio Ana Chaves, Lagoa Amélia, and Calvário (Bannerman 1915a, Naurois 1984d, J&B, Harrison & Steele 1989, Atkinson *et al* 1991). Snow (1950) found it only above 1000 m, but from there up to the Pico, where it was also found by Nadler (1993); Günther & Feiler (1985) found it only in the area of Lagoa Amélia (1500 m). Uncommon on the eastern slopes (Naurois 1972b). Usually far from human habitation (Naurois 1984d), but seen near Zampalma and São Nicolau (N–centre) by Eccles (1988). Recorded in cocoa plantations by Naurois (1984d) and (near Santa Catarina, W coast) by P.D. Alexander-Marrack (*in litt*) but not found in plantations by Correia (1928–29b) or Jones & Tye (1988). Occasionally recorded in the N, at Guadalupe, Praia das Conchas and Morro Peixe (Bocage 1904, Atkinson *et al* 1991). At least once recorded on Ilhéu das Rolas (Bocage 1889b, 1891).

   These records indicate that it is presently widespread, at least in primary and secondary forest, but patchy in distribution or easily overlooked. However, when calling it is conspicuous. It is perhaps generally less numerous than formerly (Bocage 1904, Bannerman 1915a, Naurois 1984d) although still common over much of the SW. In plantations, may not have recovered from a decline attributed to pesticide use in the early 1970s (Naurois 1984d). Considered 'Vulnerable, D1' (population <1000 adults) by BirdLife International (2000).

Habits Usually seen singly (Correia 1928–29b, J&T), it is generally found at moderate to upper levels in large trees, where it sits quietly, looking around for prey (Naurois 1984d, J&T). Song and calls loud; some described by Naurois (1984d) and Christy & Clarke (1998); also a squawking woodpecker-like call (both sexes) and a loud, down-scale, hawk-like scream *waaah* or *rerrr* (J&T). Calls and song tape-recorded (J&T, J&B, P.D. Alexander-Marrack); sonogram of one call in Nadler (1993). Sometimes calls every 7–8 sec, persistently for several minutes (J&B); males counter-sing with each other and reply to tape recordings (Nadler 1993). Stomach contents include seeds, Hemiptera and larval and adult Coleoptera (Naurois 1984d).

Breeding A nest in construction and another attributed to this species were described by Snow (1950) and Naurois (1984d) respectively. Eggs unknown. Breeding data, based on gonad condition, were summarised by Naurois (1984d) and indicate breeding at least Aug–Dec and possibly as early as Jun. Alexander thought it was breeding Jan–Feb (Bannerman 1915a). Juveniles collected Jul and Aug (Salvadori 1903b, Frade & Vieira dos Santos 1977); other juveniles and moulting adults collected or seen Dec–Mar (Naurois 1984d, Sargeant 1994).

European Golden Oriole	*Oriolus oriolus*	**VPM;–;–**
	(Linnaeus 1758)	

Port: Papa-figos-europeu (F&V).

Palearctic migrant to Africa. Vagrant Príncipe.

One immature male collected Baía de Santo António, 11 Nov 1954 (Frade 1958).

DICRURIDAE

Príncipe Drongo	*Dicrurus modestus*	**RB**;–;–**
	Hartlaub 1849	

Maria-Palu-feiticeira, Mac-Palou, Mapalu, Feiticeira (Dohrn 1866, K, Sousa 1887, B 1903, N 1994); Chota (R); Rabotizoura, Rabopeixe, Crequetché (C&C). Port: Drongo da ilha do Príncipe (N 1994).

Endemic to Príncipe; treated as conspecific with mainland *coracinus* and *atactus* (perhaps an ancient intermediate form) by Naurois (1987c), and Fry *et al* (2000) as a subspecies of *D. adsimilis* by Dowsett & Forbes-Watson (1993) and Sibley & Monroe (1990).

First recorded by Weiss (Hartlaub 1849). Declined worryingly in the early 1970s, probably owing to pesticide abuse (Naurois & Castro Antunes 1973). Appears subsequently to have recovered somewhat; in the late 1980s, frequent to common, often in pairs, in open areas with scattered trees, including Santo António town, and along roads through plantations (Jones & Tye 1988). Present, though inconspicuous, in understorey of dense second growth and plantations (J&B, Atkinson *et al* 1994). Keulemans (1866) considered it also common in the forests of the uninhabited areas. If it is not a true forest bird, it probably has a smaller population than most Príncipe endemics, because of the limited occurrence of its habitat (*cf* Snow 1950). It is currently classified as 'Near Threatened' by BirdLife International (2000).

Habits Song imitative, copying Blue-breasted Kingfisher *Halcyon malimbica*, Príncipe Glossy Starling *Lamprotornis ornatus* (Dohrn 1866, Keulemans 1866), Gulf of Guinea Thrush *Turdus olivaceofuscus* (J&B; *qv*); other calls described by Keulemans (1866), Snow (1950) and Christy & Clarke (1998); imitative subsong tape-recorded (J&B). Song from a house roof said to prophesy death (Dohrn 1866, Keulemans 1866); other superstitions attached to it were described by Keulemans (1866). According to folklore, the bird acquired its name and reputation from the sorceress Maria Palu, on whose roof it used to sit. When a priest visited her house one day, her evil spirit took fright and shot up the chimney, where it found and entered the drongo, changing the bird's colour to black with red eyes and endowing it with the powers of the witch. The inhabitants did not know what colour the bird had been before this incident (Keulemans 1866).

Eats insects, including caterpillars, butterflies, Coleoptera, Diptera, Hymenoptera, Orthoptera, mainly caught by aerial sallying (Keulemans 1866, Keulemans *in* Shelley 1896–1912, Snow 1950, J&T); also sallies to catch insects on the ground (Naurois 1987c). Reportedly fond of grasshoppers, even pursuing them into houses (Keulemans 1866, who did not record what effect this had on the inhabitants).

Breeding Dohrn (1866) believed that they bred in Sep, at the start of the rainy season, and Keulemans (1866) thought the breeding season was Oct–Jan. Gonads enlarged Sep–Dec (Naurois 1987c). Young birds accompany adults long after fledging (Keulemans 1866). Nest described by Keulemans (1866) but breeding otherwise unknown.

STURNIDAE

Chestnut-winged Starling   *Onychognathus fulgidus fulgidus*   –;RB*;–
Hartlaub 1849

Pastro (B 1891, Correia 1928–29b, R). Port: Estorninho-de-asa-castanha (F&V, N).

Subspecies endemic to São Tomé, others on Bioko and African mainland.

Very common in the late 19th century (Bocage 1891) but thought rare by Correia (1928–29a). At present common in forest, forest edge and plantations with tall trees (including coconut plantations); often seen in groups of up to 8 birds, high in trees (Günther & Feiler 1985, Jones & Tye 1988). Up to at least 850 m but not seen above this altitude in 1987 or 1988 (J&T, J&B). Not seen in the city or dry northern areas by Jones & Tye (1988), though recorded from the N by Newton (Bocage 1889b, 1891) and Sargeant (1994). Newton also collected several on Ilhéu das Rolas (Bocage 1889b, 1891).

Habits Calls varied, often abrupt, always distinctive, mostly flutey whistles and mewing sounds, eg *myeeeuw, chip-cheewoo, choo, chyoo-see, pliup, treedoo, chirrip*; often one or two call types repeated many times with successive calls separated by 3–5 sec, with slight variations; tape-recorded (J&T, J&B); sonogram of one call presented by Nadler (1993); other descriptions in Christy & Clarke (1998). One seen to swallow a large fruit, almost as large as its head; another carrying oil-palm fruit; one ate bits of *Musanga* fruit fingers (J&T). Otherwise eats small fruit, seeds and invertebrates (Naurois 1994). Birds often search clumps of epiphytic ferns on tree trunks and drink nectar from *Erythrina* (J&T).

Breeding Apparently unrecorded; a juvenile collected by Correia, Oct (Themido 1938); seen entering tree holes, and cries of juveniles heard from one hole by Naurois (1994) who thought the breeding season probably 'automne', or (Naurois 1972b) sometime between Sep and Mar.

Príncipe Glossy Starling   *Lamprotornis ornatus*   RB**;–;–
(Daudin 1800)

Torninho (Sousa 1887, not Torninko: Keulemans *in* Bannerman 1930–51); Esturninho-volgar (Correia 1928–29b); Torinho (N 1972b); Estorinho (R). Port: Estorninho do Príncipe (F&V, N 1994).

Endemic to Príncipe.

The record by Weiss in Hartlaub (1857) which Bocage (1891) took to be from São Tomé, was actually listed by Hartlaub as 'Ilha do Principe; St. Thomé', and probably referred to Príncipe (see account of Blue-breasted Kingfisher *Halcyon malimbica*). In the anonymous English translation of Hartlaub (1850), which was probably written by H.E. Strickland (Anon. 1850), it was stated that the species was found by Weiss only on Príncipe. The specimen in question, which was in ZMH, was destroyed together with the museum catalogues in World War II (C. Hinkelmann *in litt*).

Abundant to very abundant, often in flocks of up to 100, in plantations and forest regrowth throughout and in hill forests in the south (Dohrn 1866, Keulemans 1866, Snow 1950, Jones & Tye 1988); usually high in trees (Dohrn 1866, J&T). Conspicuous towards evening, when birds in the north fly high towards Pico Papagaio, where they probably roost; birds in the south undertake similar movements (Correia 1928–29b, Jones & Tye 1988; *cf* Grey Parrot *Psittacus erithacus*). From sea-level up to at least 650 m (J&T), *pace* Naurois (1972b) who affirmed that it was only found above 200 m altitude and only overlaps with *L. splendidus* at 200–300 m. Seems always to have been very common (eg Dohrn 1866, Keulemans 1866, Bannerman 1914). At least formerly hunted for food, the young birds being especially valued (Keulemans 1866).

Habits Wings make a fanning noise when birds fly (perhaps produced by notches in primaries: see Bannerman 1930–51 vol. 6); vocalisations described by Keulemans (1866), Snow (1950) and Christy & Clarke (1998); also gives a variety of twanging, gurgling and mewing notes (J&T, J&B). Calls and wing noises tape-recorded (J&T, J&B). Occasionally in mixed parties with Grey Parrots *Psittacus erithacus* or with Príncipe Golden Weaver *Ploceus princeps* and Dohrn's Thrush-babbler *Horizorhinus dohrni* (J&T). Food includes insects, spiders, snails, nestlings of Golden Weaver, bananas and berries, including drupes of *Dracaena* (Correia 1928–29b, Snow 1950, Keulemans *in* Shelley 1896–1912, Naurois 1994).

Breeding Nest, eggs and nestlings described by Keulemans (*in* Shelley 1896–1912: egg description based on accounts of local people); nests in tree holes (Naurois 1972b). Nest-building Sep (Snow 1950). Smaller flocks (2–6 birds) seen in Feb by Harrison & Steele (1989) suggest breeding then. Juveniles collected Feb and Nov (Bannerman 1914, Frade & Vieira dos Santos 1977). Said to breed Oct–May (Keulemans 1866), or end of Aug to Dec (Naurois 1972b), or 'probably Sep–Dec' (Naurois 1994), or to hatch in Jan and Feb (Dohrn 1866).

Morphometrics of a trapped bird given by Atkinson *et al* (1994).

Splendid Glossy Starling   *Lamprotornis splendidus*   (R?)B;–;–
(Vieillot 1822)

Esturninho Bo-bo (Correia 1928–29b); Estorinho (R). Port: Estorninho-esplêndido (F&V, N 1994).

Afrotropical, including Bioko (which has an endemic subspecies *lessoni*). Breeds, at least intermittently, on Príncipe; birds are indistinguishable from mainland *L. s. splendidus* (Amadon 1953).

Recorded by Dohrn (1866) and Keulemans (1866) but neither Fea nor Alexander found it (Salvadori 1903a, Bannerman 1914). Eight birds were collected in 1928 in the south and at Roça Sundi by Correia (Amadon 1953) who reported that the local people knew it to be rare and distinguished it by name from Príncipe Glossy Starling *L. ornatus* (Correia 1928–29b). Several more were taken in 1970 (Frade & Vieira dos Santos 1977), while Naurois (1972b) considered it abundant in the 1960s and 1970s. Not recorded by Jones & Tye (1988) but several reported in flocks of Príncipe Glossy Starlings *L. ornatus* in Sep, by Atkinson *et al* (1994). More recently, found regularly throughout the cultivated parts of the north and centre (Christy & Clarke 1998).

Because it is easily confused with the more common resident *L. ornatus*, it is not clear whether this species is resident. In Keulemans' time, the locals knew it and thought it was the old male of *L. ornatus* (Keulemans 1866). Naurois (1983a, 1994) considered it an irregular visitor, only occurring below 350 m, and forcing Príncipe Glossy Starling to higher altitudes during its invasions, but it is not clear on what evidence this supposition is based. It seems highly unlikely that Splendid Glossy Starling visits Príncipe in some but not all years, in large enough numbers to be noticed and to breed; a regular migration or (probably) residency, seem more likely, especially since it has never been recorded on the other two islands. The two species probably represent a double invasion (Amadon 1953) and their ecological relationships are virtually unknown. Naurois (1972b) suggested the following differences between them, but his evidence is anecdotal at best and some is contradicted by subsequent observations. Naurois (1972b) found *splendidus* in shaded plantations and around human habitation and suggested that it thus avoided competition with *ornatus*; he also suggested that the breeding seasons of the two species differed (although breeding seasons quoted by Naurois seem unreliable) and that *ornatus* ate larger fruit.

Habits Eats fruit and flying termites (Christy & Clarke 1998). Foraging behaviour and calls were described by Christy & Clarke (1998).

Breeding Two juveniles were collected on the island in Mar 1970 (Frade & Vieira dos Santos 1977); three nests with nestlings Dec (Naurois 1994); these data perhaps led to Naurois' (1972b) statement that the breeding season is Dec to Mar or Apr. Naurois (1972b) found nests in tree holes, with clutches of 2 or 3. Breeding otherwise unknown.

PLOCEIDAE

Príncipe Golden Weaver   *Ploceus princeps*   RB**;–;–
                                    (Bonaparte 1850)

Mello, Melro, Merlo (K, N 1972b, R); Camusselo (N&CA, who gave vernacular probably wrongly as Camusselo do Príncipe); Port: Tecelão do Príncipe (F&V, N 1994, C&C).

Endemic to Príncipe.
   Abundant in all habitats with trees, including plantations, forest regrowth and villages (Jones & Tye 1988), and has apparently been so at least since the mid-19th century (Keulemans 1866, Correia 1928–29b, Snow 1950). Perhaps less common in SE and in dense hill forest (Keulemans 1866, Snow 1950). Recorded on Ilhéu Caroço (Naurois 1975b).

Habits Occasionally in mixed parties with Dohrn's Thrush-babbler *Horizorhinus dohrni* and Príncipe Glossy Starling *Lamprotornis ornatus* (J&T). Song and calls described by Keulemans (1866) and Snow (1950); tape-recorded (J&T, J&B). Pokes under bark, picking bits off; occasionally sallies for aerial insects (J&T); also forages on ground (Christy & Clarke 1998). Food includes bananas, berries, chillies, other fruit, seeds, *Erythrina* nectar or pollen and insects including small beetles; probably mainly insectivorous (Keulemans 1866, Snow 1950, J&T, Atkinson *et al* 1994).

Breeding Nests and eggs described by Dohrn (1866), Keulemans (1866), Naurois (1994) and Christy & Clarke (1998); Keulemans also presented a valuable account of nest-building, incubation and other aspects of breeding behaviour. Nest-building Jan (Christy & Clarke 1998); immatures collected Feb and Mar (Bannerman 1914,

Frade & Vieira dos Santos 1977). Breeding season includes May, Jun, Aug, Sep and probably Feb (Dohrn 1866, Correia 1928–29b, Snow 1950, Jones & Tye 1988); whereas Naurois (1972b) gave breeding season (without data) as Oct–Apr and probably also Jun; Keulemans (1866) thought they bred all year, with a peak in Jul and Aug. Keulemans (1866) thought it double- or sometimes triple-brooded: pairs that nested first in Jul re-nested Sep and again in Jan; those nesting first in May having a third brood in Sep–Oct. However, it is not clear what observations Keulemans made to support these statements. The exposed nests are susceptible to storm damage, lost nests being quickly replaced (Keulemans 1866). This habit and the fact that birds remain paired all year give the impression that each pair breeds all year round, although Keulemans thought this was probably false (Keulemans 1866). Nests are robbed by Blue-breasted Kingfishers *Halcyon malimbica* and Príncipe Glossy Starlings *Lamprotornis ornatus* (Correia 1928–29b).

Morphometrics of a trapped female were given in Atkinson *et al* (1994).

Vitelline Masked Weaver   *Ploceus velatus peixotoi*   **–;RB*;–**
                                   Frade & Naurois 1964

Nario, Gungo (R); Canário (M). Port: Tecelão-de-máscara (F&V, N 1994).

Afrotropical; subspecies endemic to São Tomé. Sometimes formerly referred to *P. capitalis* (eg Bocage 1867, Salvadori 1903b, Bannerman 1915a, Amadon 1953).

Apparently rare (if genuinely recorded at all) until the 20th century, with only 2 specimens before 1949: one collected in 1861 (Bocage 1867, see Salvadori 1903b), the other at Boa Entrada by Correia in Jul 1928 (Amadon 1953, L.R. Macaulay *in litt*). The 1861 specimen may not actually refer to this species (Frade & Naurois 1964), while Correia's specimen was a very young male, referred by Amadon (1953) to the '*capitalis–melanocephalus*' group but actually a specimen of *P. velatus* (S. Keith *in litt*). In 1949 Snow (1950) found a breeding colony of what was probably this species and saw other individuals in the NE; in 1954 further specimens were obtained between the city and airport (NE) (Frade 1958). By the 1960s it was and remains common to abundant in flocks in the N and city suburbs (Frade & Naurois 1964, Naurois 1972b, 1983a, Jones & Tye 1988, Nadler 1993), and occurs up to 1100 m at Bom Sucesso (Christy & Clarke 1998). Günther & Feiler (1985) considered birds in off-season plumage in the city to be this species.

Habits Calls described by Christy & Clarke (1998) and Nadler (1993), who presented 4 sonograms. Eats seeds and fruit, and possibly nectar (Christy & Clarke 1998).

Breeding Nests and nest-sites described by Frade & Naurois (1964), Günther & Feiler (1985), Jones & Tye (1988) and Christy & Clarke (1998). Nest structure claimed to be that of *P. v. velatus*, not *P. v. vitellinus* (Frade & Naurois 1964), but we are not aware of any difference between the two, and Naurois later (1972b) wrote that the nest resembled that of these two subspecies. Eggs described by Naurois (1972b). Males seen displaying at colonies of about 10 nests, Apr (Nadler 1993) and near Praia Fernão Dias (N), Jun (Günther & Feiler 1985); more song near nests Aug (J&T). Males building in Apr (Eccles 1988), Aug (Frade & Naurois 1964) and Sep (Snow 1950). 'Nesting' Jun–Jul (Atkinson *et al* 1994). However, Naurois (1972b) gave breeding season as Aug–Apr, possibly to May. Flocks included young birds, Apr (Nadler 1993).

Morphology and systematics The lack of early records might suggest that this species is a recent (<150 years) immigrant, but the differentiation of an endemic race, although poorly marked (Naurois & Wolters 1975), suggests that it is an old-established bird which was perhaps merely overlooked by earlier collectors. However, given the efforts of Correia and others, and the relative conspicuousness of the northern, savanna birds, it seems hard to believe that if it had been common they would have missed it. The subspecific differentiation does not rule out the possibility that the bird is an introduction (Frade & Naurois 1964): cf House Sparrow *Passer domesticus* in the New World. We have not seen the specimens of this race (in MNHN and IICT) and have therefore been unable to check its validity ourselves. Morphometrics of 9 trapped birds given by Atkinson *et al* (1994).

Village Weaver	*Ploceus cucullatus* (Müller 1776)	–;RB;–

Nario (R). Port: Tecelão-de-capucho-preto (F&V, N 1994).

Afrotropical, including Bioko. Perhaps resident São Tomé. Naurois (1983a, 1994) considered the island birds to belong to the SE African *nigriceps* group; morphometrics of 9 trapped birds given by Atkinson *et al* (1994).

Possibly a recent introduction. A female was obtained by Correia in 1928 (Amadon 1953), a male by Frade in July 1954 at the airport, and another male by Naurois in Feb 1971 in the city (Frade & Vieira dos Santos 1977). Naurois (1972b, 1983a, 1994) reported breeding colonies in Jan 1971 in the northern savannas and in mangroves, palms and bushes bordering coastal lagoons. One bird attributed to this species was seen in a city park by Günther & Feiler (1985). Not seen in 1987 or 1988 (J&T, J&B) but 15 in maize at Ferreiro Governo, Jul 1990 (Atkinson *et al* 1994); one male seen by Nadler (1993).

Breeding Nests and eggs described by Naurois (1994) who gave breeding season as late Nov to Jan (no data given), whereas he earlier (1972b) gave laying period as Jan–Feb (again, no data presented).

Giant Weaver	*Ploceus grandis* (Gray 1849)	–;RB**;–

Canicela, Camicella, Camixella, Camussela (B 1888c, 1891, 1904, S 1903b, R). Port: Tecelão-grande (F&V, C&C); Tecelão-grande de São Tomé (N 1994).

Endemic to São Tomé. We concur with Moreau (1960) in considering this species a probable insular derivative of *P. cucullatus*, the earlier representative of a double invasion, rather than a derivative of *P. melanocephalus* (Naurois & Wolters 1975), although the mixture of characteristics shown by *P. grandis* prevents a confident statement.

Common, often in small groups, in plantations with tall trees, forest edge and the interior of forest regrowth, in S, W, centre and near the city (Correia 1928–29b, Naurois 1983a, Jones & Tye 1988, Sargeant 1994); uncommon in primary forest (Atkinson *et al* 1991, J&T). Mostly at low altitude (J&T, Atkinson *et al* 1994) but recorded up to 1500 m (Naurois & Wolters 1975) and recently considered fairly common near the summit of the Pico (Nadler 1993). Not a savanna bird (*contra* Amadon 1953) and records from the N are sparse: recorded in the 19th century (Bocage 1891, 1904), a juvenile collected at Guadalupe in 1970 (Frade & Vieira dos Santos 1977) and 3 solitary birds seen near Praia das Conchas in 1983 (Günther & Feiler 1985). Recorded in 19th century from Ilhéu das Rolas (Bocage 1904) but now apparently extinct there (Atkinson *et al* 1991).

Habits Song like a harsh version of that of the Village Weaver *P. cucullatus*, with the same tempo but less pure, more buzzy notes; song and calls tape-recorded (J&T, J&B); other descriptions in Christy & Clarke (1998). Food includes wild and cultivated fruit including oil-palm and pawpaw, insects including beetles, molluscs and hard seeds (Hartlaub 1857, Sharpe 1869, Bocage 1888c). Foraging behaviour described by Christy & Clarke (1998).

Breeding An egg attributed to this species by Bocage (1891, 1904) probably came from some other bird; other eggs described by Naurois (1972b) and Naurois & Wolters (1975). The nest (not previously fully described) consists of a chamber with a downward-pointing entrance tube on one side, the tube not projecting beyond the chamber (Naurois & Wolters 1975, J&T). It is placed in a low (J&T) or high (Hartlaub 1857, Eccles 1988) tree; one under coconut frond (P.D. Alexander-Marrack *in litt*); usually not suspended but enclosed in vegetation (Naurois 1994). Found breeding in May (Correia 1928–29b) though 4 birds collected Jun had poorly developed gonads (Günther & Feiler 1985); copulation Dec (Christy & Clarke 1998); nest-building Jul, Aug, Dec, Jan (J&B, P.D. Alexander-Marrack *in litt*, Christy & Clarke 1998); female entering nest, Jul (Jones & Tye 1988); immatures seen Mar–Apr, Jun–Aug, sometimes in what appeared to be family parties (Günther & Feiler 1985, Jones & Tye 1988, Christy & Clarke 1998). Naurois (1994) gave breeding season, without data, as Sep–Jan, whereas he earlier (1972b) quoted it as mid-Aug to Mar.

Body weights of 4 individuals, 62–65 g (Günther & Feiler 1985). Morphometrics of 2 trapped birds given by Atkinson *et al* (1994).

São Tomé Weaver   *Ploceus sanctithomae*   –;RB**;–
                                        (Hartlaub 1848)

Tchim-tchim-tcholó (B 1888c *et al*), Tchin-tchin-xolo (R) (onomatopoeic). Port: Tecelão-maior (F&V, signifying larger, which it is not; probably in error for *P. grandis*); Tecelão de São Tomé (N 1994, C&C).

Endemic to São Tomé. We see no good phylogenetic grounds for accepting the monotypic genus *Thomasophantes* in which this weaver has been classified previously.

    Abundant throughout in most habitats with trees, including plantations, virgin and secondary forest, isolated trees, São Tomé City; frequent in bushy and wooded habitats in the northern savannas (Snow 1950, Jones & Tye 1988, J&B).

Habits Noisy, singing and calling frequently; often in pairs, family parties or groups up to 10 (Günther & Feiler 1985, Jones & Tye 1988) or occasionally more (Christy & Clarke 1998). The song and calls were described by Snow (1950) and Christy & Clarke (1998); adult calls, song and begging call of fledglings tape-recorded (J&T, J&B, P.D. Alexander-Marrack); sonograms of calls and song presented by Nadler (1993). Behaves like a woodpecker (Alexander *in* Bannerman 1915a) or nuthatch (Snow 1950), running up and down branches and trunks, often hanging upside-down like a tit (Günther & Feiler 1985, Jones & Tye 1988); pecks pieces off dead branches (J&T); gleans from leaves (Atkinson *et al* 1991). Food includes *Erythrina* nectar, seeds, caterpillars and other insects (Bocage 1888c, Snow 1950, Jones & Tye 1988, J&T).

Breeding Nests and eggs were described by Bocage (1891, 1904), Snow (1950) and Naurois (1972b, 1994); other nests of tendrils, lined with *Platycerium* leaf skeletons (J&T). Aberrant nests were described by Fisher (2004). Naurois (1972b) gave breeding season as mid-Aug to Feb, and later (1994) as Sep–Jan, but seen nest-building, carrying food and feeding fledglings in Jul and Aug (Jones & Tye 1988), nest-building in Sep (Snow 1950), Dec and Jan (Sargeant & Alexander-Marrack 1990, Christy & Clarke

1998); begging young seen in family parties Mar, Apr, Dec, Jan (Nadler 1993, Christy & Clarke 1998). Breeding season therefore probably at least Jun–Jan. Male took greatest part in building one nest (J&T). Often parasitised by Emerald Cuckoo *Chrysococcyx cupreus* (Naurois 1972b, 1994).

Morphometrics of trapped birds were given in Atkinson *et al* (1994).

Red-headed Quelea   *Quelea erythrops*   **E(VA?);RB?;–**
                              (Hartlaub 1848)

Pardal, Pardalinha (K). Port: Pardal-de-cabeça-vermelha (F&V); Naurois (1994) gave Pardal-de-bico-vermelho but this is surely in error for *Q. quelea*.

Afrotropical. Extinct Príncipe, probably resident São Tomé.

Príncipe Apparently abundant in the mid-19th century, in plains or cassava fields in flocks of 30–80 birds, often accompanying Príncipe Golden Weavers *Ploceus princeps* and Bronze Mannikins *Lonchura cucullata* (Dohrn 1866, Keulemans 1866). Must have declined rapidly to extinction there (or emigrated?), because not found in the late 19th century by Newton, nor by Fea and subsequent collectors (Bocage 1903, Salvadori 1903a, Bannerman 1914, Frade & Vieira dos Santos 1977). Perhaps originally introduced (Moreau 1966).

São Tomé The species was first described from a São Tomé specimen collected by Weiss (Hartlaub 1848). Apparently very common in the N, at least until about 1910 (Bannerman 1915a). Newton and Alexander found it on maize plots at quite high altitude, in the Saudade/Monte Café area (Bocage 1891, 1904, Bannerman 1915a). At least one flock seen by Correia (1928–29a) in razor-grass fields. Apparently common in savanna and coconut plantations in the north in the 1960s and 70s (Naurois 1972b, Frade & Vieira dos Santos 1977). One off-season male collected and other birds attributed to this species seen in small flocks near Guadalupe and Praia das Conchas in 1983 (Günther & Feiler 1985). Not seen on São Tomé or Príncipe in Jul–Aug 1987 or 1988 by Jones & Tye (1988) and J&B, at which time males should perhaps have been in breeding dress (see below). With the great reduction in cultivated area since 1975, this species may have become less numerous. Latterly regarded as common in the north by Sargeant & Alexander-Marrack (1990) but only one flock seen (at a nest colony) by Nadler (1993). Conceivably, populations on both islands could consist entirely of semi-regular or irregular migrants from the mainland.

Habits Ate grass seeds and shoots, spiders and beetles (Keulemans 1866).

Breeding Nest described by Keulemans (1866), nest and eggs by Naurois (1994) who gave breeding season as only Nov or Dec to Feb, whereas he earlier (1972b) quoted it as Dec to Apr or May. Males on Príncipe assumed breeding dress in Jun–Jul, some as early as May; bred Jun–Sep (Keulemans 1866).

[Red-billed Quelea   *Quelea quelea*   **–;?;–**
                              (Linnaeus 1758)

Afrotropical.

São Tomé A male with 2 females, and a female, reported in mixed-species flocks in maize at Ferreiro Governo (N), Jun–Jul 1990 (Atkinson *et al* 1994). However, field notes leave some doubt as to the certainty of the identification (P.W. Atkinson *in litt*), despite its acceptance by Christy & Clarke (1998). We consider its presence on the islands unproven.]

Fire-crowned Bishop   *Euplectes hordeaceus*   –;RB;–
                                 (Linnaeus 1758)
Padé-campo (R); Pássaro-do-natal. Port: Cardeal-coroa-de-fogo (F&V).

Afrotropical. Resident São Tomé.

First recorded in 1893 (Bocage 1904, Bannerman 1915a) and perhaps introduced. Not found by Fea in 1900 and uncommon near the city in 1909 (Bannerman 1915a) but common in the north in 1928–29 (Correia 1928–29b, Amadon 1953) and the 1960s and 70s (Naurois 1972b). Mostly found in the northern savannas and city (Frade & Vieira dos Santos 1977, Naurois 1983a) and especially in tall razor-grass (Correia 1928–29a) and damp places (Naurois 1994). Recorded by Reinius (1985) and Eccles (1988) but not by Jones & Tye (1988) nor by J&B, whose visits fell in Jul and Aug when probably in eclipse plumage (*cf* Fry 1961). Found common in the north in Dec 1989 by Sargeant & Alexander-Marrack (1990). Two males seen Apr 1991 (Nadler 1993), and described as common in the N and NE by Christy & Clarke (1998).

Breeding Nests and eggs described by Naurois (1994) who gave breeding season as Dec–Jan, whereas he earlier (1972b) quoted it as Dec–Apr. However, juveniles collected Oct and Mar (Frade & Vieira dos Santos 1977). Often found breeding side by side with Golden-backed Bishop *E. aureus* (Naurois 1994).

Golden-backed Bishop   *Euplectes aureus*   –;RB;–
                                  (Gmelin 1789)
Que-blan-caná-janeilo (S 1903b, B 1904); Padé-campo-amarelho (R). Port: Tecelão-aureo (F&V, N 1994).

São Tomé and coastal Gabon to Angola (Luanda and Benguela); possibly introduced to mainland from São Tomé, or *vice versa*.

Frequent in northern savannas near farmland (Jones & Tye 1988), especially in tall razor-grass (Correia 1928–29a). Also reported in similar habitat in or near the city (Bocage 1904, Naurois 1983a) and at 300 m near Ubabudo (E) (Eccles 1988). Has perhaps never been very common (*cf* Alexander *in* Bannerman 1915a), although described as abundant by Naurois (1972b) and as the commonest species of weaver on the island by Nadler (1993) (although he found it only at Praia das Conchas), at a time when males were displaying.

Habits Calls and displays described by Christy & Clarke (1998); sonogram presented by Nadler (1993). Eats seeds and insects (Naurois 1994).

Breeding Nests and eggs described by Naurois (1972b, 1994). Naurois (1994) gave breeding season as Dec–Jan, when vegetation high. However, he earlier (1972b) gave breeding season as Dec–Apr (depending on vegetation height and therefore on rainfall), and intense display in early Apr, declining mid-Apr, with a female carrying nest material 17 Apr (Nadler 1993) indicates breeding at that time. In breeding plumage, Feb and Apr (Bannerman 1915a, Eccles 1988). Naurois (1994) described it as polygamous, with the male defending a territory by performing display-flights, and birds nesting in damp depressions. Nadler (1993) found it colonial, with groups of up to 10 nests together. Often found breeding side by side with *E. hordeaceus* (Naurois 1972b, 1994).

Morphometrics of 3 trapped birds given by Atkinson *et al* (1994).

White-winged Widowbird   *Euplectes albonotatus*   –;RB;–
                                     (Cassin 1848)

Port: Viúva-de-asa-branca (F&V, N 1994).

Afrotropical. Resident São Tomé.

   First found by Alexander in 1909 (Bannerman 1930–51 vol. 7); Frade's 1954 record
was thus not the first, *contra* Naurois (1972b, 1994). Now frequent in northern savannas
near farmland and the city (Günther & Feiler 1985, Jones & Tye 1988, Sargeant 1994).

Habits Calls and displays described by Christy & Clarke (1998).

Breeding Nests and eggs described by Naurois (1972b, 1994); nests in shorter
vegetation than *E. hordeaceus* and *E. aureus* according to Naurois (1994), although he
earlier (1972b) wrote that all three nested in the same grasslands, at the same time,
each male defending its territory containing the nests and females against males of
the other species. Males mainly in non-breeding dress Jul–Aug (Jones & Tye 1988);
full breeding plumage, Apr and Dec (Eccles 1988, Nadler 1993, Sargeant 1994). Breeds
Dec–Jan according to Naurois (1994) but end Nov to Mar according to Naurois (1972b),
and breeding plumage suggests season extends at least until Apr (*cf* Golden-backed
Bishop *E. aureus*).

ESTRILDIDAE

Chestnut-breasted Negro-finch   *Nigrita bicolor*   RB;–;–
                                     (Hartlaub 1844)

Siwie-gigoe (K: NB Dutch pronunciation). Port: Negrinha (N 1994).

Afrotropical. Resident Príncipe.

   Has apparently always been uncommon (Dohrn 1866, Keulemans 1866, Bannerman
1914). Collected by Newton (Bocage 1903: only one specimen) and Alexander
(Bannerman 1914) and at Terreiro Velho (E) and 'mangroves' in 1970 (Frade & Vieira
dos Santos 1977). Naurois (1983a) considered it 'plutôt rare' in the early 1970s in
secondary forest and plantations, and perhaps less rare on Ilhéu Caroço (Naurois
1972b). One seen at Porto Real (formerly Esperança) in 1988 (J&B); one in Santo
António (Atkinson *et al* 1994); 2 between Santo António and Bela Vista, 1991 (Sargeant
1994). Described by Keulemans (1866) as shy, usually found in pairs in rank vegetation
along streams. At least formerly, occurred up to 550 m or more (Bannerman 1914).

Habits Food consists of insects, although seeds also eaten (Keulemans 1866). Song
described by Keulemans (1866) and Christy & Clarke (1998).

Breeding Nest and eggs described by Naurois (1994) but it is not clear whether the
description comes from Príncipe. Naurois (1972b) found 'occupied but empty' nests
in Jan and Mar. Keulemans (1866) thought them paired the whole year round, but
stated that they bred in the rainy season, ie Nov onward; Naurois (1994) thought
them breeding possibly Dec–Feb.

[Neumann's Waxbill   *Estrilda thomensis*   –;?;–
                                (Sousa 1888)

Western Angola. Early records from São Tomé probably erroneous or escaped cage birds.

This mainland species (see Hall & Moreau 1970) was originally described from an adult male specimen (now in Coimbra Museum) said to have been collected by Moller in 1885 on São Tomé (Sousa 1888), but it is doubtful that the specimen came from there (Bannerman 1915a, *cf* Salvadori 1903b). A second specimen was obtained at Guadalupe (Bocage 1904); Bannerman (1930–51 vol.7) gave the year as 1887, though it is unclear from where he got this date, as none was mentioned in the earlier literature. This seems to have been a genuine record, but it seems most likely that it was a vagrant or escapee, rather than a resident (Bannerman 1915a). There are no other records. **]**

Common Waxbill   *Estrilda astrild*   RB;RB;–
                            (Linnaeus 1758)

Boca-vermelha (ST only: K); Tueli (B 1879); Januário (B 1891, 1904); Bico-de-lacre (Correia 1928–29b, F&V, also Port: N 1994); Quebra-cana (N 1972b); Queblan-canan-vermelho (R). Port: Bico-de-lacre (N 1994).

Afrotropical including Bioko. Resident São Tomé and Príncipe; endemic subspecies '*sousae*', formerly recognised as having doubtful validity (Amadon 1953, Naurois 1983a) is rejected; size and plumage of specimens from the islands are within the range of variation of *angolensis* and *rubriventris* from Angola and Gabon respectively, which 2 races should probably also be united (AT unpubl.). Snow's group collected only one bird, not 'specimens' as referred to by Bannerman (1930–51, vol. 8).

Príncipe Frequent along roadsides, in gardens and farmland in the N and centre (Jones & Tye 1988, Christy & Clarke 1998). Keulemans (1866) found it uncommon and only in the south, and little known to the inhabitants (no local name there). Apparently always less common than on São Tomé (see Keulemans 1866, Bannerman 1914, Amadon 1953, Frade & Vieira dos Santos 1977) and may have become extinct between 1865 and 1949 (during which period it was unrecorded, see Bocage 1903, Snow 1950), becoming re-established sometime thereafter, perhaps by reintroduction.

São Tomé First recorded by Keulemans (1866), though many later authors overlooked his notes from São Tomé (eg Salvadori 1903b, Günther & Feiler 1985). At present abundant, usually in groups of 2–25 but up to 200 (Günther & Feiler 1985), in northern savannas and farmland, city suburbs, oil-palm and coffee plantations, grasslands, roadsides and gardens throughout (Bocage 1891, Jones & Tye 1988, Atkinson *et al* 1994); found in sugar cane by Correia (1928–29b), in the crater of Lagoa Amélia at 1400 m, which is completely surrounded by primary forest (Snow 1950), and on tracks inside cocoa plantations up to 1000 m (Eccles 1988), and in a grassy clearing surrounded by forest at Rio Quija (Nadler 1993).

Habits Sometimes in mixed flocks with Bronze Mannikins *Lonchura cucullata* and other granivores (Keulemans 1866, Snow 1950, Günther & Feiler 1985, Nadler 1993, Atkinson *et al* 1994). Recorded smoke-bathing by Correia (1928–29b). Song a quiet, buzzy, repeated *chi-chi bo derzz* (J&T); other calls described by Christy & Clarke (1998). At least formerly, caught on São Tomé (and probably Príncipe) for the bird trade (Fry 1961).

Breeding Nests isolated or in loose groups; nests and eggs described by Naurois (1994). Naurois (1972b) reported breeding on São Tomé Nov–Feb, whereas Naurois (1994) reported scattered nesting throughout the year, but without giving data; 2 fledglings seen together, Apr (Eccles 1988).

Southern Cordon-bleu   *Uraeginthus angolensis*   –;RB;–
                                   (Linnaeus 1758)

Peito-celeste (Correia 1928–29b, F; also Port: N 1994); Suin-suin (R).

Afrotropical. Resident São Tomé, perhaps introduced by man.

First recorded by Correia in 1928, who found it 'quite rare', and always in pairs (Correia 1928–29b, Amadon 1953), at localities including Praia das Conchas (Themido 1938). One specimen from São Tomé at MNCN, prepared by P. Curats in 1943, was probably a cage bird taken to Bioko (J. Pérez del Val *in litt*). Snow (1950) saw only a single flock and Naurois (1983a) considered it 'plutôt rare', but Günther & Feiler (1985) found it more common in 1983. It is now common to abundant in the city, northern savannas, and cocoa plantations at low altitude in the north and centre, usually in pairs or small parties (Eccles 1988, Jones & Tye 1988), so has probably increased since the early 1970s. Apparently absent from wetter areas (Jones & Tye 1988), though recorded as far south as Água Izé (Frade & Vieira dos Santos 1977).

Habits Sometimes in mixed flocks with other granivores (Atkinson *et al* 1994). Calls described by Christy & Clarke (1998). At least formerly caught for the bird trade (Fry 1961).

Breeding Nests and eggs described by Naurois (1994) who gave breeding season (without data) as Nov–Jan, whereas Naurois (1972b) quoted it as Jan–Mar.

Morphometrics of 2 trapped birds given by Atkinson *et al* (1994).

Bronze Mannikin   *Lonchura cucullata*   RB;RB;RB
                             (Swainson 1837)

STP: Bico-preto (N 1972b). P: Sibi-singa (K). ST: Freirinha (B 1891, 1904, Correia 1928–29b, F&V); Queblan-canan-preto (R). Port: Freirinha (N 1994).

Afrotropical including Bioko (*L. c. cucullata*). Resident Príncipe and São Tomé and Annobón.

Príncipe Common since at least the mid-19th century and currently locally abundant, in towns, roadsides and open, grassy areas in plantations and elsewhere (Hartlaub 1850, Keulemans 1866, Jones & Tye 1988). Found by Keulemans (1866) in 'plains', grass and cassava fields, in flocks of up to several 100 birds. In contrast, Correia obtained only 3 in 1928 (Amadon 1953).

São Tomé Locally abundant in similar habitats as on Príncipe, including the city and savannas (Jones & Tye 1988), sometimes together with Common Waxbills *Estrilda astrild* (Günther & Feiler 1985). On W coast of São Tomé, not seen further S than Neves by Jones & Tye (1988), though occurs in wetter areas in SE and recorded formerly at São Miguel (SW) (Bocage 1891, 1904). Occurs at least up to 800 m near Saudade and Zampalma (Bocage 1904, J&T).

Annobón First reported, as a recently arrived breeding species, in 2000 (Pérez del Val 2001).

Habits Usually in small groups or family parties of up to 20 adults and young (Jones & Tye 1988); sometimes in mixed flocks with other granivores (Atkinson *et al* 1994).

Breeding Breeds at least Jan–Feb and Apr–Oct (Dohrn 1866, Bannerman 1915a, Snow 1950, Frade 1958, Günther & Feiler 1985, Jones & Tye 1988), or Sep–Jan, perhaps extending to Apr (Naurois 1972b), or possibly all year (Keulemans 1866, Naurois 1994). Reportedly triple-brooded, with a 6-week interval between nests (Keulemans 1866). Nest, nest-sites and eggs described by Dohrn (1866), Keulemans (1866), Bocage (1891) and Naurois (1994); one nest in a street-lamp in São Tomé harbour (Günther & Feiler 1985). Incubation period 12 days (Keulemans 1866). Young reportedly eaten by Blue-breasted Kingfishers *Halcyon malimbica* (Keulemans 1866).

VIDUIDAE

Pin-tailed Whydah   *Vidua macroura*   ?;RB;–
                                          (Pallas 1764)

Viuva (B 1891, 1904); Viuvinha (R). Port: Viuvinha cauda-de-fio (F&V); Viúva-de-cauda-de-agulha (N 1994).

Afrotropical including Bioko. Resident São Tomé, no certain records Príncipe. Not recorded Annobón (*contra* Naurois 1994).

Príncipe Reported only by Naurois (1983a) who termed it 'rare', without further comment. Correia's (1929) dark blue or black 'Viuva', which he collected on Príncipe 'the same' as he collected on Bioko, appears to refer to Príncipe Drongo *Dicrurus modestus* (L.R. Macaulay *in litt*).

São Tomé Frequent in open areas in SE near Santa Cruz, Ribeira Peixe, Rio Martim Mendes, Porto Alegre (Günther & Feiler 1985, Jones & Tye 1988, Christy & Clarke 1998); common in the N (Bocage 1889c, Eccles 1988, Nadler 1993), though not seen there by J&T. In the past, occurred in the highlands up to Trás-os-Montes and Lagoa Amélia (Bannerman 1915a, Snow 1950, Frade & Vieira dos Santos 1977); at present, at least up to Trindade (260 m) (Eccles 1988). Formerly recorded in the city (Snow 1950), SW and extreme S (Salvadori 1903b, Bocage 1904, Frade & Vieira dos Santos 1977) and perhaps more widely distributed than at present; very abundant in savannas, clearings, roadsides, gardens and plantations in the early 1970s according to Naurois (1972b). In the past, at least, caught for the bird trade (Fry 1961).

Habits Often in large groups, up to *c* 30 birds, probably polygamous, and sometimes accompanies flocks of Bronze Mannikins *Lonchura cucullata* (Jones & Tye 1988).

Breeding During Jun–Sep, most birds were in 'female'-type plumage, though some males were in breeding dress (Snow 1950, Günther & Feiler 1985, Jones & Tye 1988). Males in breeding plumage singing in Sep and Dec (Snow 1950, Sargeant 1994). Eggs found Nov–Jan (Naurois 1994); the egg is described by Bocage (1891) and Naurois (1994). Probably parasitises Bronze Mannikin and possibly Common Waxbill *Estrilda astrild* and Vitelline Masked Weaver *Ploceus velatus* (Naurois 1972b, 1994).

Long-tailed Paradise-Whydah   *Vidua paradisaea*   –;?;–
                                    (Linnaeus 1766)

Afrotropical. Probable escaped birds, São Tomé.

Recorded twice, one collected by Moller at Nova Moca, 800 m (Bocage 1887c) and an adult male in full breeding plumage shot by Alexander, Feb 1909 (Bannerman 1915a). Moller's specimen showed evidence of having been in captivity (Bocage 1887c, 1904) though Alexander's did not, and Bannerman (1915a) speculated that it was a breeding resident or vagrant from the mainland. However, since no other records have been reported, it seems likely that both examples were escaped cage birds, or that the second was a vagrant.

FRINGILLIDAE

Yellow-fronted Canary   *Serinus mozambicus*   –;RB;?
                             (Müller 1776)

Canário (R). Port: Xirico de São Tomé (for 'endemic' race: F&V); Canário de Moçambique (N 1994).

Afrotropical. Resident São Tomé; vagrant or escaped cage bird Annobón. We do not accept as valid the endemic race '*santhome*' of Bannerman (1921, 1930–51 vol. 6): the type and other specimens examined (including the specimen from Annobón) fall within the range of variation of mainland *tando* (AT unpubl., Naurois 1975b, *cf* Amadon 1953).

São Tomé Abundant in city and its suburbs, feeding in casuarinas and oil-palms; frequent in farmland in northern savannas; apparently absent from wetter areas further S (Naurois 1975b, Jones & Tye 1988), though recorded at Nova Moca in the highlands (Frade & Vieira dos Santos 1977). Apparently rather rare in the 19th to mid-20th centuries (eg Bocage 1904, Correia 1928–29b, Naurois 1972b) and found by Correia only in the 'Iron trees' (*Margaritoria discoidea*?) whose seeds it ate, along the N coast. At least formerly, captured as a cage bird (Correia 1928–29b) or for the bird trade (Fry 1961) and still commonly kept as a pet (Snow 1950, J&T); may have been introduced by man (Salvadori 1903b).

Annobón The single record, a male collected in Aug 1959 (Fry 1961), was probably a vagrant from the mainland or an escaped cage bird.

Breeding Sings much Dec–Jan (Christy & Clarke 1998); nest-building, Apr (Eccles 1988). No other breeding data; all specimens collected had regressed gonads (Naurois 1975b).

Morphometrics of 4 trapped birds given by Atkinson *et al* (1994).

Príncipe Seedeater   *Serinus rufobrunneus*   RB**
                          (Gray 1862)   **(São Tomé and Príncipe)**

Padé, Pardal (the former a corruption of the latter) (K, B 1888c *et al*, N 1975b). Port: Canário-castanho de São Tomé e Príncipe, Bigodinho de São Tomé e Príncipe (N 1994).

Endemic to São Tomé and Príncipe, with three subspecies (see Naurois 1975b). Often placed in endemic genus *Poliospiza*.

Serinus rufobrunneus rufobrunneus   **RB*;–;–**
(Gray 1862)

Additional names: Chota-café (C&C). Port: Chamariço do Príncipe (F&V).

Subspecies endemic to Príncipe; others on Ilhéu Caroço and São Tomé.

This race was considered rare and restricted to bushy uncultivated places on the W coast by Dohrn (1866) and Keulemans (1866: 375), though Keulemans (1866: 394) described it as very common in the west, occurring singly or in pairs, in the bushes of the coast, as well as being one of the commonest birds in the forests of the uninhabited parts. The true status perhaps lay between the two (*cf* Naurois 1975b). At that time, the coastal vegetation was secondary bush, an early stage in the regeneration of forest, and probably contained many seeds which the birds could eat (Naurois 1975b). Fea found it in the SE as well as the W (Salvadori 1903a) and Alexander found it close to the peak (*cf* his collecting dates and itinerary in Bannerman 1914). Correia collected several more in 1928–29 (Amadon 1953). It seems still to have been common in 1949 in a variety of habitats including coconut plantations and forest edge (Snow 1950). Naurois (1975b) found it uncommon in 1970 but more common in 1972 and 1973. Jones & Tye (1988) saw only 2 together, in coconut palm crowns over a clearing at Terreiro Velho (E). It is reportedly common, however, on the narrow southern coastal fringe where there are still coconut palms (J. Baillie, M. Dallimer & M. Melo, pers comm). Considering the range of habitats occupied by the São Tomé subspecies and the apparent lack of competitors on Príncipe, the rarity and restricted occurrence of this noisy (see Snow 1950) and conspicuous bird is presently inexplicable. There must be grounds for concern over its long-term survival.

Habits Vocalisations were described by Snow (1950) and Naurois (1975b). Food includes berries, seeds, insects and vegetable matter (Keulemans 1866, Snow 1950, Naurois 1975b). Sometimes behaves like a nuthatch *Sitta* running up and down trunks and large branches (Naurois 1975b).

Breeding A nest attributed to this species was described by Naurois (1975b). Birds with inactive gonads or otherwise showing no sign of reproductive activity in Mar, Jun, Jul and Sep; birds with active gonads or otherwise showing breeding activity in Mar, Jul, Nov and Dec, although data for intervening periods are lacking (Snow 1950, Naurois 1975b; *cf P. r. thomensis*).

Serinus rufobrunneus thomensis   **–;RB*;–**
(Bocage 1888a)

Additional name: Port: Chamariço de São Tomé (F&V).

Subspecies endemic to São Tomé; others on Príncipe and Ilhéu Caroço.

Fea found this race 'common' in 1900 (Salvadori 1903b) but Alexander considered it uncommon in 1909 (Bannerman 1915a). Correia (1928–29b) found it not common but generally distributed. Naurois considered it 'peu commun' or 'local ... on peut cheminer une ou deux heures dans les plantations ombragées ou la Capoeira (forêt secondaire) sans l'apercevoir ....' (Naurois 1972b, 1975b) or alternatively 'abondant, forêts; alt. basses et moyennes' (Naurois 1983a), all describing the same period, the early 1960s and 1970s. Anyhow, seems to have increased since then, as it is now one of the most frequently encountered birds in forested country, rendered conspicuous by its loud (*contra* Naurois 1975b) calls. It is common to abundant in all habitats with trees, including city suburbs, southern forests, plantations and northern dry woodland; probably most numerous in forest and regrowth in S and in mountains, least common in N (Jones & Tye 1988). Occurs commonly up to at least 1500 m (Naurois 1975b,

Günther & Feiler 1985) and seen near the summit of the Pico by Nadler (1993). At least formerly occurred on Ilhéu das Rolas (Bocage 1904) but now apparently extinct there (Atkinson *et al* 1991).

Habits Sometimes associates with São Tomé Speirops *Speirops lugubris* (Günther & Feiler 1985). Occurs high in tall trees, low in undergrowth, and frequently seen on the ground (Naurois 1975b, J&T). Calls were described by Snow (1950) and Naurois (1975b), some reminiscent of Nightingale *Luscinia megarhynchos* (J&T); song includes whistles and bubbling trills (J&B); calls and song tape-recorded (J&T, J&B, P.D. Alexander-Marrack); Nadler (1993) presented sonograms of call and distress call (in the hand). Stomach contents include seeds; once seen poking into pawpaw fruits *Carica papaya* (Naurois 1975b) and seen eating *Musanga* fruit (J&B). Gleans insects from twigs, leaves and rocks in streams; creeps up trees, probing into bark (J&T). Once seen plucking 3–4-cm long leaves of an unknown plant species, running them back and forth through the bill, mandibulating their petioles, sometimes holding them by a foot or picking the leaf to pieces, and apparently swallowing most of it (J&T).

Nest and eggs, unknown until the 1960s, were described by Naurois (1972b, 1975b). Data summarised by Naurois (1994, see also Günther & Feiler 1985) indicate a breeding season at least extending from May to Jan, with a break in Sep. Birds with inactive gonads have been collected in all months of the year except Oct. A lull in breeding activity may coincide with the period of heaviest rainfall in Feb–May, similarly a Sep lull would coincide with the lesser rainfall peak in that month (Naurois 1975b).

Morphometrics of a trapped bird were given in Atkinson *et al* (1994).

Serinus rufobrunneus fradei   **RB*;–;–**
(Naurois 1975b)

Additional name: Port: Chamariço do Jóquei (F&V).

Subspecies endemic to Ilhéu Caroço (Ilhéu Boné de Jóquei) off southern tip of Príncipe (Plate 28). The islet measures only 35 ha, rises to 300 m and is separated from Príncipe by 3 km of shallow sea (max. depth 50 m). The 2 were probably joined 10,000–15,000 years ago. We have not been able to check the validity of this subspecies, whose only specimens are at MNHN. Other subspecies on Príncipe and São Tomé.

Abundant in the early 1970s, with a density estimated at 30–40 ha^{-1}, giving a total population of 150–200 in the oil-palm *Elaeis guineensis* forest which is its known habitat; possibly more live in other parts of the islet, which has no other common land-bird (Naurois 1975b); still very abundant in 2004 (M. Melo pers comm). Considered 'Endangered' by Gascoigne (1995).

Habits Stomach contents included pulp of oil-palm fruits and macerated seeds (Naurois 1975b). Calls incessantly; other aspects of behaviour and morphology were presented by Naurois (1975b).

Breeding Of 6 adults collected in Nov, one male had active gonads and one female was about to lay; 4 other females had inactive gonads (Naurois 1975b). These and other observations suggested breeding Jun–Aug and Nov (Naurois 1975b). A nest was described by Naurois (1975b).

São Tomé Grosbeak   *Neospiza concolor*   –;RB***;–
                               (Bocage 1888d)

Enjoló (B 1888d).

Monotypic genus endemic to São Tomé. Originally classed as a grosbeak-weaver (Ploceidae: Bocage 1888d, see also Moreau 1960), it is probably a true finch (Fringillidae), close to Príncipe Seedeater *Serinus rufobrunneus* (Salvadori 1903b, Bocage 1904, Bannerman 1930–51 vol. 7, Amadon 1953, 1965, Naurois 1975b, 1988c). Indeed, it and *S. rufobrunneus* probably represent a double invasion, the latter coming later; if so, the two are sister groups within the genus *Serinus*, and the genus *Neospiza* should be submerged within the former on phylogenetic grounds.

Discovered by Newton, who collected a male at Angolares (E) in Jul 1888 (Bocage 1888d) and 2 males in the São Miguel-Rio Quija area (SW) in Sep 1890 (Bocage 1891, 1904); the precise locality for these 2 is unrecorded (Naurois 1988c), although they were taken in forest (Bocage 1891). Two specimens were probably destroyed in the Bocage Museum fire of 1978; the remaining one, the type, from Rio Quija, is in BMNH (Naurois 1983a, 1988c). Naurois (1988c) gave photographs of one of the lost specimens. It was sought fruitlessly in the same area by Correia (1928–29a,b), Naurois (1975b) and Atkinson *et al* (1991) but rediscovered in August 1991 above the Rio Xufexufe (SW) (Sargeant *et al* 1992): on two days, one bird was seen on a bare tree in a tree-fall gap in closed-canopy primary forest, on a ridge at 230m altitude. An adult and a juvenile were seen in the same area in Aug 1997 (I. Sinclair *per* A. Stattersfield *in litt*), with another brief probable sighting, again in the same area, in Sep 1997 (P. Kaestner *in litt*). One bird seen in Jan 2002 at 380 m on a ridge above the Rio São Miguel and another photogrpahed on 9 Feb 2002 on a ridge at 400 m in the same area where more than one bird was probably present (Dallimer *et al* 2003). A female was captured in Jan 2003 near Monte Carmo (Ribeira Piexe) and 2 males captured Feb 2005 in same area as Dallimer's observations (M. Melo pers comm). Considered 'Critically Endangered, D1' (population <50 adults) by BirdLife International (2000), despite the current lack of information on its population size and distribution.

Habits The voice was described by Sargeant *et al* (1992) as a brief series of 4–5 short, thin, canary-like whistles, and by Dallimer *et al* (2003) as a 2-note whistle, with the second note higher, frequently repeated, recalling Príncipe Seedeater but deeper in tone, simpler, and more repetitive. Described as having 'an odd, jerky flight' (P. Kaestner *in litt*). Feeds on fruits of *Uapaca guineensis* and *Dicranolepis thomensis* (Dallimer *et al* 2003).

APPENDIX 1

Summary checklist of the birds of the Gulf of Guinea islands

R = resident (R* = endemic subspecies; R** = endemic species; R*** = endemic genus); B = breeds (seabirds only); M = migrant (Afrotropical or Palearctic); V = vagrant / regular migrant in small numbers; E = extinct; I = introduced; ? = status uncertain. Offshore records are shown only where the species is not certainly associated with a particular island. Data for Bioko from Pérez del Val (1996) and Pérez del Val *et al* (1997).

	Bioko	Príncipe	São Tomé	Annobón	offshore
PROCELLARIIDAE					
Bulwer's Petrel *Bulweria bulwerii*	-	?	?	-	?
Cory's Shearwater *Calonectris diomedea*	V	-	-	-	?
Great Shearwater *Puffinus gravis*	-	-	-	-	?
Sooty Shearwater *P. griseus*	V	?	?	V	?
HYDROBATIDAE					
Wilson's Storm-Petrel *Oceanites oceanicus*	-	-	-	-	V
Black-bellied Storm-Petrel *Fregetta tropica*	-	-	-	-	V
British Storm-Petrel *Hydrobates pelagicus*	V	?	?	-	?
Madeiran Storm-Petrel *Oceanodroma castro*	-	B?	B?	?	M
Leach's Storm-Petrel *O. leucorhoa*	-	-	-	-	V
PODICIPEDIDAE					
Little Grebe *Tachybaptus ruficollis*	R	-	-	-	
PHAETHONTIDAE					
Red-billed Tropicbird *Phaethon aethereus*	-	?	?	-	
White-tailed Tropicbird *P. lepturus*	V	B	B	B	
SULIDAE					
Cape Gannet *Sula capensis*	M	?	?	-	
Masked Booby *S. dactylatra*	-	V?	-	V?	V
Red-footed Booby *S. sula*	-	?	-	-	?
Brown Booby *S. leucogaster*	V	B	B	B?	
PHALACROCORACIDAE					
White-breasted Cormorant *Phalacrocorax carbo*	-	?	?	-	
Reed Cormorant *P. africanus*	V	-	R		
ANHINGIDAE					
Darter *Anhinga rufa*	V	-	-	-	
FREGATIDAE					
Ascension Frigatebird *Fregata aquila*	-	?	-	-	?
ARDEIDAE					
Little Bittern *Ixobrychus minutus*	-	R?	-	-	
Black-crowned Night Heron *Nycticorax nycticorax*	-	-	?	-	
Squacco Heron *Ardeola ralloides*	V	?	-	-	
Cattle Egret *Bubulcus ibis*	M	R	R	M?	
Green-backed Heron *Butorides striatus*	R	R	R	V	
Black Heron *Egretta ardesiaca*	-	-	V	-	
Western Reef Heron *E. gularis*	R	R	R	R	
Little Egret *E. garzetta*	M	?	?	-	
Great White Egret *Ardea alba*	-	-	?	-	

	Bioko	Príncipe	São Tomé	Annobón	offshore
Purple Heron *A. purpurea*	-	-	V	-	
Grey Heron *A. cinerea*	V	V	V	-	
CICONIIDAE					
Yellow-billed Stork *Mycteria ibis*	-	-	V	-	
White Stork *Ciconia ciconia*	-	-	V	-	
THRESKIORNITHIDAE					
Hadada *Bostrychia hagedash*	R	-	-	-	
Olive Ibis *B. olivacea*	-	R*	-	-	
Dwarf Ibis *B. bocagei*	-	-	R**	-	
PHOENICOPTERIDAE					
Greater Flamingo *Phoenicopterus roseus*	-	-	?	-	
Lesser Flamingo *Phoeniconaias minor*	-	V	-	-	
ANATIDAE					
Knob-billed Duck *Sarkidiornis melanotos*	-	-	V	-	
ACCIPITRIDAE					
Honey Buzzard *Pernis apivorus*	V	-	-	-	
Black (Yellow-billed) Kite *Milvus migrans*	M	R	R	V	
African Fish Eagle *Haliaeetus vocifer*	V	-	-	-	
Palm-nut Vulture *Gypohierax angolensis*	R	-	-	-	
African Goshawk *Accipiter tachiro*	R*	-	-	-	
FALCONIDAE					
Common Kestrel *Falco tinnunculus*	-	?	-	-	
Western Red-footed Falcon *F. vespertinus*	-	-	V	-	
Peregrine Falcon *F. peregrinus*	V	-	V	-	
PHASIANIDAE					
Feral Chicken *Gallus gallus*	-	-	-	R	
Helmeted Guineafowl *Numida meleagris*	E	?	R	R	
Harlequin Quail *Coturnix delegorguei*	-	-	R*	-	
Red-necked Spurfowl *Francolinus afer*	-	-	R	-	
RALLIDAE					
Buff-spotted Flufftail *Sarothrura elegans*	R	-	-	-	
African Crake *Crex egregia*	V	V	V	-	
African Water Rail *Rallus caerulescens*	-	-	V	-	
Allen's Gallinule *Porphyrio alleni*	M	V	V?	V	
Common Moorhen *Gallinula chloropus*	-	R	R	E	
Lesser Moorhen *G. angulata*	V	V	R?	-	
HELIORNITHIDAE					
African Finfoot *Podica senegalensis*	R?	-	-	-	
GLAREOLIDAE					
Black-winged Pratincole *Glareola nordmanni*	-	V	-	V	
Rock Pratincole *G. nuchalis*	V	-	-	-	
CHARADRIIDAE					
Little Ringed Plover *Charadrius dubius*	V	-	?	-	
Ringed Plover *C. hiaticula*	M	-	?	-	
White-fronted Sand Plover *C. marginatus*	V	-	V	-	
'Lesser' Golden Plover *Pluvialis sp.*	-	-	V	-	

	Bioko	Príncipe	São Tomé	Annobón	offshore
Eurasian Golden Plover *P. apricaria*	-	-	?	-	
Grey Plover *P. squatarola*	V	-	V		
SCOLOPACIDAE					
Sanderling *Calidris alba*	-	-	-	M	
[Little] Stint *C. [minuta]*	-	?	-	-	
Pectoral Sandpiper *C. melanotos*	-	V	-	-	
Curlew Sandpiper *C. ferruginea*	V	V	V	-	
Common Snipe *Gallinago gallinago*	V	-	-	-	
Bar-tailed Godwit *Limosa lapponica*	-	V	V	V	
Whimbrel *Numenius phaeopus*	V	M	M	M	
Eurasian Curlew *N. arquata*	-	V	V	-	
Common Redshank *Tringa totanus*	V	-	-	-	
Marsh Sandpiper *T. stagnatilis*	M	-	-	-	
Greenshank *T. nebularia*	V	M	M	-	
Green Sandpiper *T. ochropus*	V	-	?	-	
Wood Sandpiper *T. glareola*	V	-	M	-	
Common Sandpiper *Actitis hypoleucos*	M	M	M	-	
Ruddy Turnstone *Arenaria interpres*	-	V	V	-	
STERCORARIIDAE					
Pomarine Skua *Stercorarius pomarinus*	-	-	V	-	
Arctic Skua *S. parasiticus*	-	?	?	?	?
Long-tailed Skua *S. longicaudus*	-	?	?	-	?
LARIDAE					
Lesser Black-backed Gull *Larus fuscus*	-	-	-	-	?
Sabine's Gull *Larus sabini*	-	?	?	-	?
Royal Tern *Sterna maxima*	V	?	?	-	?
Sandwich Tern *S. sandvicensis*	-	-	M	-	
Common Tern *S. hirundo*	V	?	M	-	
Arctic Tern *S. paradisaea*	V	?	?	-	
Bridled Tern *S. anaethetus*	-	B?	B?	B	
Sooty Tern *S. fuscata*	V	B	B?	B?	
Whiskered Tern *Chlidonias hybrida*	V	-	-	-	
Black Tern *C. niger*	M	?	?	-	?
White-winged Black Tern *C. leucopterus*	-	-	-	V	
Black Noddy *Anous minutus*	-	B	B?	B	
Common Noddy *A. stolidus*	V	B	B	B	
COLUMBIDAE					
African Green Pigeon *Treron calva*	R	R*	-	-	
São Tomé Green Pigeon *T. sanctithomae*	-	-	R**	-	
Tambourine Dove *Turtur tympanistria*	R	-	-	-	
São Tomé Bronze-naped Pigeon *Columba malherbii*	-	R**	R**	R**	
Lemon Dove *C. larvata*	R*	R*	R*	R*	
Olive (Rameron) Pigeon *C. arquatrix*	R	-	-	-	
Maroon Pigeon *C. thomensis*	-	-	R**	-	
Feral Pigeon *C. livia*	-	R	R	-	
Red-eyed Dove *Streptopelia semitorquata*	R	-	-	-	
Laughing Dove *S. senegalensis*	-	R	R	-	
PSITTACIDAE					
Grey Parrot *Psittacus erithacus*	R	R	R?	-	
Red-headed Lovebird *Agapornis pullaria*	E/I?	E	R	-	

	Bioko	Príncipe	São Tomé	Annobón	offshore
MUSOPHAGIDAE					
Great Blue Turaco *Corythaeola cristata*	R	-	-	-	
Verreaux's Turaco *Turaco macrorhynchus*	R	-	-	-	
CUCULIDAE					
Jacobin Cuckoo *Oxylophus jacobinus*	-	V	V	-	
Striped Cuckoo *O. levaillantii*	V	-	-	-	
Great Spotted Cuckoo *Clamator glandarius*	-	?	-	V	
European/African Cuckoo *Cuculus canorus/gularis*	-	-	V	-	
Red-chested Cuckoo *Cuculus solitarius*	R(*)	-	-	-	
Emerald Cuckoo *Chrysococcyx cupreus*	R	R*	R*	R(*?)	
Klaas's Cuckoo *C. klaas*	R	-	R?	-	
Didric Cuckoo *C. caprius*	R	-	-	-	
Green Coucal *Ceuthmochares aereus*	R	-	-	-	
TYTONIDAE					
Barn Owl *Tyto alba*	R	-	R*	-	
STRIGIDAE					
Annobón Scops Owl *Otus scops*	-	-	-	R*	
Príncipe Scops Owl *Otus* sp	-	?	-	-	
São Tomé Scops Owl *O. hartlaubi*	-	-	R**	-	
Fraser's Eagle Owl *Bubo poensis*	R	-	-	-	
Wood Owl *Strix woodfordii*	R	-	-	-	
CAPRIMULGIDAE					
Pennant-winged Nightjar *Macrodipteryx vexillaria*	M	-	-	-	
APODIDAE					
São Tomé Spinetail *Zoonavena thomensis*	-	R**	R**	-	
Sabine's Spinetail *Rhaphidura sabini*	R	-	-	-	
Mottled Spinetail *Telacanthura ussheri*	V	-	-	-	
Cassin's Spinetail *Neafrapus cassini*	R?	-	-	-	
Scarce Swift *Schoutedenapus myoptilus*	R*	-	-	-	
Palm Swift *Cypsiurus parvus*	R	R	R	-	
Black Swift *Apus barbatus*	V?	?	?	-	
Common Swift *A. apus*	-	-	?	-	
Little Swift *A. affinis*	R*	R*	R*	-	
Alpine Swift *Tachymarptis melba*	-	-	-	M?	
TROGONIDAE					
Bar-tailed Trogon *Apaloderma vittatum*	R	-	-	-	
ALCEDINIDAE					
Chocolate-backed Kingfisher *Halcyon badia*	R	-	-	-	
Blue-breasted Kingfisher *H. malimbica*	-	R*	-	-	
Senegal Kingfisher *H. senegalensis*	M	-	-	-	
White-bellied Kingfisher *Corythornis leucogaster*	R*	R*	-	-	
Malachite Kingfisher *C. cristata*	-	-	R*	-	
Giant Kingfisher *Megaceryle maxima*	V	-	-	-	
Pied Kingfisher *Ceryle rudis*	R	?(B)	V	-	
MEROPIDAE					
Blue-headed Bee-eater *Merops muelleri*	R*	-	-	-	
CORACIIDAE					
European Roller *Coracias garrulus*	-	V	V	-	
Blue-throated Roller *Eurystomus gularis*	R	-	-	-	

	Bioko	Príncipe	São Tomé	Annobón offshore
BUCEROTIDAE				
Black-casqued Wattled Hornbill *Ceratogymna atrata*	R	-	-	
CAPITONIDAE				
Speckled Tinkerbird *Pogoniulus scolopaceus*	R*	-	-	-
Yellow-throated Tinkerbird *P. subsulphureus*	R*	-	-	-
Yellow-rumped Tinkerbird *P. bilineatus*	R*	-	-	-
INDICATORIDAE				
Least Honeyguide *Indicator exilis*	R*	-	-	-
PICIDAE				
Tullberg's Woodpecker *Campethera tullbergi*	R	-	-	-
Buff-spotted Woodpecker *C. nivosa*	R*	-	-	-
Elliot's Woodpecker *Dendropicos elliotii*	R	-	-	-
EURYLAIMIDAE				
Grey-headed Broadbill *Smithornis sharpei*	R*	-	-	-
HIRUNDINIDAE				
Mountain Saw-wing *Psalidoprocne fuliginosa*	R	-	-	-
Banded Martin *Riparia cincta*	-	V	-	-
Grey-rumped Swallow *Pseudhirundo griseopyga*	-	?	-	-
Lesser Striped Swallow *Hirundo abyssinica*	R	-	-	-
Barn Swallow *H. rustica*	M	V	M	-
House Martin *Delichon urbicum*	-	V	-	-
MOTACILLIDAE				
Yellow Wagtail *Motacilla flava*	M	-	-	-
Mountain Wagtail *M. clara*	R	-	-	-
Wagtail *Motacilla* sp	-	?	-	-
Tree Pipit *Anthus trivialis*	M	-	?	-
CAMPEPHAGIDAE				
Grey Cuckoo-shrike *Coracina caesia*	R	-	-	-
PYCNONOTIDAE				
Mountain Greenbul *Andropadus tephrolaemus*	R	-	-	-
Little Greenbul *A. virens*	R*	-	-	-
Cameroon Sombre Greenbul *A. curvirostris*	R	-	-	-
Slender-billed Greenbul *A. gracilirostris*	R	-	-	-
Yellow-whiskered Greenbul *A. latirostris*	R	-	-	-
Golden Greenbul *Calyptocichla serina*	R	-	-	-
Cameroon Olive Greenbul *Phyllastrephus poensis*	R	-	-	-
Icterine Greenbul *P. icterinus*	R	-	-	-
Green-tailed Bristlebill *Bleda eximia*	R	-	-	-
Red-tailed Greenbul *Criniger calurus*	R	-	-	-
TURDIDAE				
Forest Robin *Stiphrornis erythrothorax*	R	-	-	-
Bocage's Akalat *Sheppardia bocagei*	R*	-	-	-
White-bellied Robin-Chat *Cossyphicula roberti*	R	-	-	-
Fire-crested Alethe *Alethe diademata*	R	-	-	-
Brown-chested Alethe *A. poliocephala*	R	-	-	-
White-tailed Ant-Thrush *Neocossyphus poensis*	R	-	-	-
Rufous Flycatcher-Thrush *N. fraseri*	R*	-	-	-
Stonechat *Saxicola torquata*	R	-	-	-
Whinchat *S. rubetra*	-	V	-	-

	Bioko	Príncipe	São Tomé	Annobón	offshore
Gulf of Guinea Thrush *Turdus olivaceofuscus*	-	R**	R**	-	
African Thrush *T. pelios*	R*	-	-	-	
SYLVIIDAE					
Evergreen Forest Warbler *Bradypterus lopezi*	R*	-	-	-	
Great Reed Warbler *Acrocephalus arundinaceus*	-	-	V	-	
Greater Swamp Warbler *A. rufescens*	R	-	-	-	
São Tomé Prinia *Prinia molleri*	-	-	R**	-	
Green Longtail *Urolais epichlora*	R*	-	-	-	
Black-capped Apalis *Apalis nigriceps*	R	-	-	-	
Buff-throated Apalis *A. rufogularis*	R	-	-	-	
Grey Apalis *A. cinerea*	R	-	-	-	
White-tailed Warbler *Poliolais lopezi*	R*	-	-	-	
Yellow-browed Camaroptera *Camaroptera superciliaris*	R*	-	-	-	
Olive-green Camaroptera *C. chloronota*	R*	-	-	-	
Yellow Longbill *Macrosphenus flavicans*	R	-	-	-	
Grey Longbill *M. concolor*	R	-	-	-	
Rufous-crowned Eremomela *Eremomela badiceps*	R	-	-	-	
Willow Warbler *Phylloscopus trochilus*	V	-	-	-	at sea
Wood Warbler *P. sibilatrix*	V	-	-	-	
Black-capped Woodland Warbler *P. herberti*	R*	-	-	-	
Garden Warbler *Sylvia borin*	V	-	V	V	
Green Hylia *Hylia prasina*	R	-	-	-	
Bocage's Longbill *Amaurocichla bocagii*	-	-	R***	-	
MUSCICAPIDAE					
Fraser's Forest-Flycatcher *Fraseria ocreata*	R	-	-	-	
Spotted Flycatcher *Muscicapa striata*	V	V	V	V	at sea
African Dusky Flycatcher *M. adusta*	R*	-	-	-	
Yellow-footed Flycatcher *M. sethsmithi*	R	-	-	-	
MONARCHIDAE					
White-bellied Crested Flycatcher *Elminia albiventris*	R	-	-	-	
São Tomé Paradise Flycatcher *Terpsiphone atrochalybeia*	-	-	R**	-	
Red-bellied Paradise Flycatcher *T. rufiventer*	R*	-	-	-	
Annobón Paradise Flycatcher *T. smithii*	-	-	-	R**	
PLATYSTEIRIDAE					
Shrike-Flycatcher *Megabyas flammulatus*	R	-	-	-	
Chestnut Wattle-eye *Dyaphorophyia castanea*	R	-	-	-	
Black-necked Wattle-eye *D. blissetti (chalybea)*	R	-	-	-	
Bioko Batis *Batis poensis*	R*	-	-	-	
PICATHARTIDAE					
Grey-necked Picathartes *Picathartes oreas*	R	-	-	-	
TIMALIIDAE					
Pale-breasted Illadopsis *Illadopsis rufipennis*	R*	-	-	-	
Blackcap Illadopsis *I. cleaveri*	R*	-	-	-	
Grey-chested Illadopsis *Kakamega poliothorax*	R	-	-	-	
African Hill Babbler *Pseudoalcippe abyssinica*	R	-	-	-	
Dohrn's Thrush-babbler *Horizorhinus dohrni*	-	R***	-	-	
REMIZIDAE					
Tit-hylia *Pholidornis rushiae*	R*	-	-	-	

	Bioko	Príncipe	São Tomé	Annobón	offshore
NECTARINIIDAE					
Yellow-chinned Sunbird *Anthreptes rectirostris*	R*	-	-	-	
Little Green Sunbird *A. seimundi*	R*	-	-	-	
Fraser's Sunbird *Deleornis fraseri*	R*	-	-	-	
Príncipe (Hartlaub's) Sunbird *Anabathmis hartlaubii*	-	R**	-	-	
São Tomé (Newton's) Sunbird *A. newtonii*	-	-	R**	-	
Giant Sunbird *Dreptes thomensis*	-	-	R***	-	
Collared Sunbird *Hedydipna collaris*	R*	-	-	-	
Blue-throated Brown Sunbird *Cyanomitra cyanolaema*	R*	-	-	-	
Cameroon Blue-headed Sunbird *C. oritis*	R*	-	-	-	
Olive Sunbird *C. olivacea*	R	R	-	-	
Green-throated Sunbird *Chalcomitra rubescens*	R*	-	-	-	
Olive-bellied Sunbird *Cinnyris chloropygia*	R	-	-	-	
Tiny Sunbird *C. minulla*	R*	-	-	-	
Northern Double-collared Sunbird *C. reichenowi*	R*	-	-	-	
Bates's Sunbird *C. batesi*	R	-	-	-	
Ursula's Mouse-coloured Sunbird *C. ursulae*	R	-	-	-	
ZOSTEROPIDAE					
Yellow White-eye *Zosterops senegalensis*	R*	-	-	-	
Príncipe White-eye *Z. ficedulinus*	-	R**	R**	-	
Annobón White-eye *Z. griseovirescens*	-	-	-	R**	
São Tomé Speirops *Speirops lugubris*	-	-	R**	-	
Príncipe Speirops *S. leucophaeus*	-	R**	-	-	
Fernando Po Speirops *S. brunneus*	R**	-	-	-	
LANIIDAE					
São Tomé (Newton's) Fiscal *Lanius newtoni*	-	-	R**	-	
Lesser Grey Shrike *L. minor*	-	V	-	V	
Red-backed Shrike *L. collurio*	-	-	V	V	
MALACONOTIDAE					
Fülleborn's Black Boubou *Laniarius fuelleborni*	R*	-	-	-	
ORIOLIDAE					
Black-winged Oriole *Oriolus nigripennis*	R	-	-	-	
São Tomé Oriole *O. crassirostris*	-	-	R**	-	
European Golden Oriole *O. oriolus*	V	V	-	-	
DICRURIDAE					
Fork-tailed Drongo *Dicrurus adsimilis*	R	-	-	-	
Príncipe Drongo *D. modestus*	-	R**	-	-	
CORVIDAE					
Pied Crow *Corvus albus*	R	-	-	-	
STURNIDAE					
Narrow-tailed Starling *Poeoptera lugubris*	V	-	-	-	
Chestnut-winged Starling *Onychognathus fulgidus*	R	-	R*	-	
Waller's Red-winged Starling *O. walleri*	R	-	-	-	
Príncipe Glossy Starling *Lamprotornis ornatus*	-	R**	-	-	
Splendid Glossy Starling *L. splendidus*	R*	R?	-	-	
PASSERIDAE					
Grey-headed Sparrow *Passer griseus*	R	-	-	-	

	Bioko	Príncipe	São Tomé	Annobón	offshore
PLOCEIDAE					
Black-necked Weaver *Ploceus nigricollis*	R*	-	-	-	
Black-billed Weaver *P. melanogaster*	R	-	-	-	
Príncipe Golden Weaver *P. princeps*	-	R**	-	-	
Vitelline Masked Weaver *P. velatus*	-	-	R*	-	
Village Weaver *P. cucullatus*	R	-	R	-	
Giant Weaver *P. grandis*	-	-	R**	-	
Maxwell's Black Weaver *P. albinucha*	R*	-	-	-	
Dark-backed Weaver *P. bicolor*	R	-	-	-	
Brown-capped Weaver *P. insignis*	R	-	-	-	
São Tomé Weaver *P. sanctithomae*	-	-	R**	-	
Red-headed Malimbe *Malimbus rubricollis*	R*	-	-	-	
Red-headed Quelea *Quelea erythrops*	V	E(V?)	R?	-	
Red-billed Quelea *Q. quelea*	-	-	?	-	
Fire-crowned Bishop *Euplectes hordeaceus*	-	-	R	-	
Golden-backed Bishop *E. aureus*	-	-	R	-	
Yellow Bishop *E. capensis*	R	-	-	-	
White-winged Widowbird *E. albonotatus*	-	-	R	-	
ESTRILDIDAE					
Grey-crowned Negro-finch *Nigrita canicapilla*	R	-	-	-	
Pale-fronted Negro-finch *N. luteifrons*	R*	-	-	-	
Chestnut-breasted Negro-finch *N. bicolor*	-	R	-	-	
White-breasted Negro-finch *N. fusconota*	R	-	-	-	
Little Olive-back *Nesocharis shelleyi*	R	-	-	-	
Red-faced Crimsonwing *Cryptospiza reichenowii*	R	-	-	-	
Green Twinspot *Mandingoa nitidula*	R*	-	-	-	
Neumann's Waxbill *Estrilda thomensis*	-	-	?	-	
Common Waxbill *E. astrild*	R	R	R	-	
Black-crowned Waxbill *E. nonnula*	R*	-	-	-	
Southern Cordon-bleu *Uraeginthus angolensis*	-	-	R	-	
Bronze Mannikin *Lonchura cucullata*	R	R	R	R	
Black and White Mannikin *L. bicolor*	R	-	-	-	
VIDUIDAE					
Pin-tailed Whydah *Vidua macroura*	R	?	R	-	
Long-tailed Paradise-Whydah *V. paradisaea*	-	-	?	-	
FRINGILLIDAE					
Yellow-fronted Canary *Serinus mozambicus*	-	-	R		?
Príncipe Seedeater *S. rufobrunneus*	-	R**	R**	-	
São Tomé Grosbeak *Neospiza concolor*	-	-	R***	-	
Oriole Finch *Linurgus olivaceus*	R	-	-	-	

APPENDIX 2

Gazetteer

Co-ordinates for rivers refer to their mouths or, if tributaries, to their downstream confluences. The word 'Roça' (estate) is omitted from most former estate names. Names and co-ordinates are taken from the 1:25,000 *Carta de S Tomé* and *Carta de Príncipe* (Ministério do Ultramar 1958, 1962); longitudes on these maps are now known to be in error but corrections are not yet available. Place-names and co-ordinates for Annobón are taken from the 1:40,000 *Mapa de la República de Guinea Ecuatorial* (IGN 1986); additional Annobonese place-names are taken from Basilio (1957) and Loboch (1962).

Annobón

A Bobo, Río (= San Juan)	NE	1° 25'S 5° 39'E
Adams, Isla de (= Ye Cayi)	<1 km S of SE tip	1° 26'S 5° 38'E
Ajabal, Bahía	E coast	1° 26'S 5° 39'E
Angandyi, Bahía	E coast	1° 25'S 5° 39'E
A Pot, Lago (crater lake, A-poto *or* Mazafin)	N centre	1° 25'S 5° 38'E
Áquequel (= Aqueque)	NW	1° 25'S 5° 37'E
Ate, Bahía	E coast	1° 27'S 5° 39'E
Aual (*also* Bahía Aual)	W coast	1° 26'S 5° 37'E
Capuchinos, Montaña de (= Quepuchin)	NE	1° 25'S 5° 39'E
crater lake (Lago A Pot, A-poto *or* Mazafin)	N centre	1° 25'S 5° 38'E
Cus	centre	1° 26'S 5° 38'E
Escobár, Isla de (= Ycögö, Ye Fogo)	3–4 km S of SW tip	1° 29'S 5° 38'E
Estephania, Pico	[untraced; = Santamina?]	
Fernando Póo, Isla de (= Ye Hum, Ye Mun)	3–4 km S of SW tip	1° 29'S 5° 38'E
Fogo, Pico do *or* Pico del Fuego	N	1° 25'S 5° 38'E
Jobo Sopa, Punta	W coast	1° 25'S 5° 37'E
Manjab, Punta	S coast	1° 27'S 5° 37'E
Mofina, Punta	S coast	1° 27'S 5° 38'E
Monanbayu, Punta	W coast	1° 27'S 5° 37'E
Palé (= Palea, San António de Palea)	N coast	1° 24'S 5° 38'E
Quepuchin (= Capuchinos, Montaña de)	NE	1° 25'S 5° 39'E
Quioveo (Pico)	centre	1° 26'S 5° 38'E
San Antonio (Mábana)	S coast	1° 27'S 5° 38'E
Sandegadu, Punta	W coast	1° 24'S 5° 37'E
San Juan, Río (= A Bobo)	NE	1° 25'S 5° 39'E
San Pedro (de Agany *or* Angandyi *or* Anganchi)	E	1° 25'S 5° 39'E
Santamina (= Macizo Santamina)	SE	1° 27'S 5° 38'E
Santarem, Isla de (= Ye Cátretu, Ye Cachiselu)	3–4 km S of SW tip	1° 29'S 5° 38'E
Tortuga, Isla (= Ye Ygany, Yegany, Ye Yngandyi)	1.5 km off NE coast	1° 24'S 5° 39'E

Príncipe

Agulhas, Baía das (= Bahía do Oeste)	W	1° 37'N 7° 21'E
airport	N centre	1° 40'N 7° 24'E
Banzul, Rio (= Ribeira Banzú)	W centre	1° 37'N 7° 23'E
Bela Vista	NE	1° 37'N 7° 25'E
Bom-bom, Ilhéu	just off NW tip	1° 42'N 7° 24'E
Boné de Jóquei, Ilhéu (= Ilhéu Caroço)	3 km SE of SE tip	1° 31'N 7° 26'E
Bonézinho, Ilhéu	off Ilhéu Caroço	1° 31'N 7° 26'E
Caroço, Ilhéu (= Ilhéu Boné de Jóquei)	3 km SE of SE tip	1° 31'N 7° 26'E
Carriote (= Cariote)	SW	1° 35'N 7° 22'E
Esperança (= Porto Real)	centre	1° 37'N 7° 24'E
Galé, Pedra da	4km NW of Ilhéu Bom-bom	1° 44'N 7° 23'E
Infante Dom Henrique	SE	1° 34'N 7° 25'E

Izé, Ribeira	NW	1°41′N 7°23′E
Maria Correia	SW	1°36′N 7°21′E
Mencorne, Picos	SE	1°35′N 7°24′E
Mesa, A	SW	1°35′N 7°21′E
Mosteiros, Ilhéus dos	just off NE tip	1°41′N 7°28′E
Negro, Pico	S tip	1°32′N 7°24′E
Oeste, Bahía do (= Baía das Agulhas)	W	1°37′N 7°21′E
Oeste, Roça (= São Joaquim)	W	1°37′N 7°23′E
Ôque Pipi	E	1°36′N 7°25′E
Papagaio, Pico	centre	1°37′N 7°24′E
Papagaio, Rio	centre E	1°38′N 7°25′E
Portinho, Ilhéus	S coast	1°32′N 7°23′E
Porto Real (= Esperança)	centre	1°37′N 7°24′E
Praia Seca	S coast	1°33′N 7°24′E
Príncipe, Pico do	S centre	1°35′N 7°23′E
Ribeira Fria	E coast	1°35′N 7°25′E
Santana, Ilhéu	NE coast	1°40′N 7°27′E
Santo António	NE	1°38′N 7°25′E
Santo António, Baía de	NE	1°39′N 7°26′E
São Joaquim (= Roça Oeste)	W	1°37′N 7°23′E
São Tomé, Ribeira de	SW	1°34′N 7°21′E
Sundi	NW	1°40′N 7°23′E
Terreiro Velho	E	1°37′N 7°25′E
Tinhosa Grande, Ilhéu (Ilhas Tinhosas)	25 km SSW of Príncipe	1°21′N 7°18′E
Tinhosa Pequeno, Ilhéu (Ilhas Tinhosas)	21 km SSW of Príncipe	1°23′N 7°18′E
Tinhosas, Ilhas (Pedras)	21–25 km SSW of Príncipe	1°22′N 7°18′E

São Tomé

Abade, Rio	E	0°14′N 6°43′E
Afonso, Ribeira	E coast	0°11′N 6°42′E
Agulha, Ponta da	E coast	0°15′N 6°45′E
airport	NW	0°23′N 6°43′E
Alegre, Porto	extreme S	0°02′N 6°32′E
Almeirim	NE	0°19′N 6°43′E
Amélia, Lagoa	N centre	0°17′N 6°36′E
Ana Chaves, Rio	S centre	0°06′N 6°38′E
Angolares (São João dos) = Santa Cruz	SE coast	0°08′N 6°39′E
Angra Toldo, Rio	SE	0°09′N 6°41′E
As Almas (= 'Zalma')	NE	0°18′N 6°45′E
Bindá	W coast	0°13′N 6°28′E
Boa Entrada	N	0°21′N 6°40′E
Boca do Inferno	SE near Agua Izé	0°13′N 6°44′E
Bombaim	centre	0°15′N 6°38′E
Bom Sucesso	centre	0°17′N 6°37′E
Budo-tap'ana	[untraced]	
Cabras, Ilhéu das	just off NE coast	0°24′N 6°43′E
Café, Monte	N centre	0°18′N 6°39′E
Calvário	centre	0°16′N 6°35′E
Cão Grande	S centre	0°07′N 6°34′E
Carregado, Morro	N coast	0°24′N 6°37′E
Cascata	W	0°18′N 6°34′E
Caué, Rio	SE	0°05′N 6°35′E
Chamiço	NW	0°19′N 6°36′E
Conchas, Praia das	N coast	0°24′N 6°37′E
Contador, Rio	NW	0°21′N 6°32′E
Cruzeiro	SE	0°11′N 6°38′E
Diogo Nunes (*and* Praia Diogo Nunes)	NE	0°23′N 6°42′E
Ermelinda	S centre	0°08′N 6°35′E
Esperança, Morro	NW centre	0°17′N 6°36′E
Fernão Dias (*and* Praia Fernão Dias)	N coast	0°24′N 6°41′E

Ferreiro Governo	NE	0° 22'N 6° 42'E
Formoso Grande, Pico	centre	0° 13'N 6° 38'E
Formoso Pequeno, Pico	centre	0° 14'N 6° 37'E
Fortunato	NW	0° 19'N 6° 35'E
Gentio, Morro	SE	0° 09'N 6° 38'E
Granja, Roça	SE	0° 08'N 6° 38'E
Guadalupe	N	0° 23'N 6° 38'E
Guadalupe, Rio	N	0° 24'N 6° 38'E
Iógo-Iógo (Enseada do)	extreme S coast	0° 02'N 6° 32'E
Ió Grande, Rio	SE	0° 06'N 6° 38'E
Izé, Agua (= Ribeira Izé)	E coast	0° 13'N 6° 44'E
Jou, Roça	SW coast	0° 07'N 6° 30'E
Juliana de Sousa	W coast	0° 12'N 6° 28'E
Lembá	W coast	0° 15'N 6° 26'E
Lembá, Rio	W	0° 15'N 6° 28'E
Malanza, Rio and Laguna de	extreme S	0° 03'N 6° 32'E
Martim Mendes, Rio	SE	0° 06'N 6° 38'E
Melão, Praia	NE coast	0° 19'N 6° 45'E
Mendes da Silva	N	0° 13'N 6° 41'E
Mestre António	N	0° 15'N 6° 44'E
Micoló (and Praia do)	NE coast	0° 24'N 6° 42'E
Minho, Monte	N centre	0° 16'N 6° 38'E
Mogadinho	[untraced]	
Moinho, Dependência	N	0° 20'N 6° 39'E
Monte Carmo	S	0° 08'N 6° 35'E
Mussucavú, Rio	SW	0° 06'N 6° 31'E
Neves	NW coast	0° 22'N 6° 33'E
Nova Moca	N centre	0° 17'N 6° 38'E
Ouro, Rio do	NE	0° 24'N 6° 41'E
Palma, Ribeira	NW	0° 21'N 6° 35'E
Pedroma	NE	0° 16'N 6° 42'E
Peixe, Morro	N coast	0° 24'N 6° 39'E
Peixe, Ribeira (= Perseverança)	SE coast	0° 05'N 6° 37'E
Pico (de São Tomé)	NW	0° 16'N 6° 33'E
Pinheira	N	0° 17'N 6° 43'E
Quija, Rio	SW	0° 07'N 6° 30'E
Quimpo	N	0° 15'N 6° 43'E
Quixibá, Ilhéu	just off SE coast	0° 04'N 6° 35'E
Rolas, Ilhéu das	2 km off S tip	0° 00'N 6° 31'E
Santa Catarina	W coast	0° 16'N 6° 28'E
Santa Cruz (= Angolares)	SE coast	0° 08'N 6° 39'E
Santana	E coast	0° 15'N 6° 45'E
Santana, Ilhéu de	just off E coast	0° 14'N 6° 46'E
Santo Amaro, Campo de	N near airport	0° 22'N 6° 42'E
Santo Antônio	SW coast	0° 06'N 6° 31'E
São João, Rio	SE	0° 08'N 6° 39'E
São João dos Angolares (= Santa Cruz)	SE coast	0° 08'N 6° 39'E
São Miguel	SW coast	0° 08'N 6° 29'E
São Nicolau	centre	0° 17'N 6° 38'E
São Tomé city	NE	0° 20'N 6° 44'E
Saudade, Roça	N centre	0° 17'N 6° 38'E
Sete Pedras	5 km off SE coast	0° 02'N 6° 37'E
Tamarindos, Praia dos	N coast	0° 24'N 6° 39'E
Tomé, Agua	trib. Rio Abade	0° 14'N 6° 43'E
Tras-os-Montes	centre	0° 16'N 6° 36'E
Trindade	central NE	0° 18'N 6° 41'E
Triunfo	[untraced: N savannas]	
Ubabudo	E	0° 16'N 6° 42'E
Xufexufe, Rio	SW	0° 07'N 6° 30'E
Zampalma (and Zampalma Velha)	centre	0° 16'N 6° 37'E

BIBLIOGRAPHY

Early volumes of *Ibis* are numbered as, for example, (10)6, meaning series 10 volume 6. A similar system is followed for the *Jornal de Sciencias Mathematicas, Physicas e Naturaes, Lisboa*, whose numbering systems for volumes and parts were independent, leading to confusion if only one is cited. For this periodical (and some others), (2)1(3) means series 2, volume 1, part 3.

References marked with an asterisk are not specifically cited in the text but are included for completeness; we have tried to include all important works dealing with the islands' ornithology, especially early ones and those in little-known or discontinued periodicals, and including all references that give first descriptions of endemic taxa. References to descriptions of non-endemic taxa are not listed here but may be traced from the authority given in the Systematic List and in Peters (1931–86). For birds, this bibliography is more complete than that of Gascoigne (1993, 1996), which also contains some errors in publication dates, but the latter should be consulted for references dealing with other taxa, and for additional bird references dealing with Bioko (Fernando Póo). A botanical bibliography has been published by Figueiredo (1994d).

Alexander-Marrack, P.D. 1990. Unpubl. report on a visit to São Tomé, 18–22 August 1990.

Allen, W. & Thomson, T.R.H. 1848. *A Narrative of the Expedition sent by Her Majesty's Government to the River Niger in 1841 under the Command of Captain H.D. Trotter.* Richard Bentley, London.

Amadon, D. 1953. Avian systematics and evolution in the Gulf of Guinea. *Bull. Am. Mus. Nat. Hist.* 100: 393–452.

Amadon, D. 1965. Position of the genus *Neospiza* Salvadori. *Ibis* 107: 395–396.

Amadon, D. & Short, L.L. 1992. Taxonomy of lower categories – suggested guidelines. *Bull. Br. Orn. Club* 112A: 11–38.

Anon. [H.E. Strickland?] 1850. Ornithology of the coasts and islands of western Africa, by Dr. Hartlaub. *Jardine's Contrib. Orn.* 1850: 129–140.

Anon. 1994. Tour d'horizon: le programme ECOFAC. *Canopée* 1: 2–3.

Atkinson, I.A.E. 1985. The spread of commensal species of *Rattus* to oceanic islands and their effects on island avifaunas. *In* Moores, P.J. (ed) *Conservation of Island Birds.* pp. 35–81. Tech. Publ. No. 3. International Council for Bird Preservation, Cambridge.

Atkinson, P., Peet, N. & Alexander, J. 1991. The status and conservation of the endemic bird species of São Tomé and Príncipe, West Africa. *Bird Conserv. Internat.* 1: 255–282.

Atkinson, P.W., Dutton, J.S., Peet, N.B. & Sequeira, V. (eds) 1994. *A Study of the Birds, Small Mammals, Turtles and Medicinal Plants of São Tomé with notes on Príncipe.* Study Report 56, BirdLife International, Cambridge.

Bannerman, D.A. 1912. On a collection of birds made by Mr. Willoughby P. Lowe on the west coast of Africa and outlying islands; with field-notes by the collector. *Ibis* (9)6: 219–268.

Bannerman, D.A. 1914. Report on the birds collected by the late Mr. Boyd Alexander (Rifle Brigade) during his last expedition to Africa. Part I. The birds of Prince's Island. *Ibis* (10)2: 596–631.

Bannerman, D.A. 1915a. Report on the birds collected by the late Mr. Boyd Alexander (Rifle Brigade) on his last expedition to Africa. Part II. The birds of St. Thomas' Island. *Ibis* (10)3: 89–121.

Bannerman, D.A. 1915b. Report on the birds collected by the late Mr. Boyd Alexander (Rifle Brigade) during his last expedition to Africa. Part III. The birds of Annobón Island. *Ibis* (10)3: 227–234.

Bannerman, D.A. 1919. [Untitled: ibises from Cameroon and Príncipe.] *Bull. Br. Orn. Club* 40: 4–7.

Bannerman, D.A. 1921. *Serinus mozambicus santhomé*, subsp. nov. *Bull. Br. Orn. Club* 41: 137.

*Bannerman, D.A. 1922. On the Emerald and Golden Cuckoos of Africa. *Novit. Zool.* 29: 413–420.

Bannerman, D.A. 1930–51. *The Birds of Tropical West Africa* Crown Agents, London.

*Bannerman, D.A. 1931. The Maroon Pigeon of São Thomé. *Ibis* (13)1: 652–654.

Barns, T.A. 1928. In Portuguese West Africa: Angola and the isles of the Guinea Gulf. *Geogr. J.* 72: 18–37.

Barrena, N. 1911. La isla de Annobón. Un paso para su conocimiento. *La Guinea Española*, Revista quincenal publicada por los Misioneros Hijos del Inmaculada Corazón de María, Sta Isabel, Fernando Po. [Article published in instalments, in issues 13 of 10 Jul 1909 to 12 of 25 Jun 1911. Birds in 1911 issues 6: 46–47 (25 March) and 7: 52 (10 April).]

*Barrena, N. 1925–26. La Isla de Annobón. Su corografía su etnología y la Misión de los Hijos del Inmaculado Corazón de María. Por un Misionero del Vicariato de Fernando Póo, 1924. *La Guinea Española* 589ff. [2nd edn of above, with corrections. Not seen in original but may not include faunal notes; see Barrena & Perramón 1965.]

*Barrena, N. & Perramón, R. 1965. *La Isla de Annobón*. Publ. Inst. Claretiano de Africanistas de la Misión Católica de Santa Isabel, Secc. Monogr. I, Barcelona. [Collected articles of Barrena 1925–26, with notes and additions by Perramón. Does not include bird notes of Barrena 1911.]

Basilio, A. 1957. *Caza y Pesca en Annobón. Aves de la Isla. La Pesca de la Ballena* Instituto de Estudios Africanos, Madrid.

Basilio, A. 1963. *Las Aves de la Isla de Fernando Poo*. COCULSA, Madrid.

BDPA 1985. *Potencialidades agrícolas: República Democrática de São Tome e Príncipe*. Bureau pour le Développement de la Production Agricole, Paris.

Bibby, C.J., Collar, N.J., Crosby, M.J., Heath, M.F., Imboden, C., Johnson, T.H., Long, A.J., Stattersfield, A.J. & Thirgood, S.J. 1992. *Putting Biodiversity on the Map: Priority areas for global conservation*. International Council for Bird Preservation, Cambridge.

BirdLife International 2000. *Threatened Birds of the World*. Lynx Edicions, Barcelona and BirdLife International, Cambridge.

Bocage, J.V.B. du 1867. Aves das possessões portuguezas da Africa occidental que existem no Museu de Lisboa. *J. Sci., math. phys. nat. Lisboa* 1(2): 129–153.

Bocage, J.V.B. du 1879. Subsidios para a fauna das possessões portuguezas d'Africa occidental. *J. Sci. math. phys. nat. Lisboa* 7(26): 85–96.

* Bocage, J.V.B. du 1880. Aves de Bolama e da Ilha do Príncipe. *J. Sci. math. phys. nat. Lisboa* 8(29): 71–72.

Bocage, J.V.B. du 1887a. Oiseaux nouveaux de l'île St. Thomé. *J. Sci. math. phys. nat. Lisboa* 11(44): 250–253.

* Bocage, J.V.B. du 1887b. Additamento á fauna ornithologica de S. Thomé. *J. Sci. math. phys. nat. Lisboa* 12(46): 81–83.

Bocage, J.V.B. du 1887c. Lista das aves de S. Thomé colligidas pelo sr. Moller em 1885. *Inst. Rev. Sci. Lit. Coimbra* (2)34: 562 565.

Bocage, J.V.B. du 1888a. Sur un oiseau nouveau de St. Thomé de la famille Fringillidae. *J. Sci. math. phys. nat. Lisboa* 12(47): 148–150.

* Bocage, J.V.B. du 1888b. Note sur la "*Phaeospiza thomensis*". *J. Sci. math. phys. nat. Lisboa* 12(47): 192.

Bocage, J.V.B. du 1888c. Sur quelques oiseaux de l'île St. Thomé. *J. Sci. math. phys. nat. Lisboa* 12(48): 211–215.

Bocage, J.V.B. du 1888d. Oiseaux nouveaux de l'île St. Thomé. *J. Sci. math. phys. nat. Lisboa* 12(48): 229–232.

Bocage, J.V.B. du 1889a. Breves considerações sobre a fauna de S. Thomé. *J. Sci. math. phys. nat. Lisboa* (2)1(1): 33–36.

Bocage, J.V.B. du 1889b. Sur deux espèces à ajoutter [*sic*] à la faune ornithologique de St. Thomé. *J. Sci. math. phys. nat. Lisboa* (2)1(2): 142–144.

Bocage, J.V.B. du 1889c. Aves da Ilha de S. Thomé. *J. Sci. math. phys. nat. Lisboa* (2)1(3): 209–210.

Bocage, J.V.B. du 1891. Oiseaux de l'île St. Thomé. *J. Sci. math. phys. nat. Lisboa* (2)2(6): 77–87.

Bocage, J.V.B. du 1893a. Note sur deux oiseaux nouveaux de l'île Anno-Bom. *J. Sci. math. phys. nat. Lisboa* (2)3: 17–18.

Bocage, J.V.B. du 1893b. Mamiferos, aves e reptis da Ilha de Anno-Bom. *J. Sci. math. phys. nat. Lisboa* (2)3(9): 43–46.

* Bocage, J.V.B. du 1896. Aves d'Africa de que existem no Museu de Lisboa os exemplares typicos. *J. Sci. math. phys. nat. Lisboa* (2)4(15): 179–186.

Bocage, J.V.B. du 1903. Contribution à la faune des quatre îles du Golfe de Guinée. *J. Sci. math. phys. nat. Lisboa* (2)7(25): 25–59. [In three parts, dealing with Bioko, Príncipe, Annobón.]

Bocage, J.V.B. du 1904. Contribution à la faune des quatre îles du Golfe de Guinée. IV. Ile de St. Thomé. *J. Sci. math. phys. nat. Lisboa* (2)7(26): 65–96.

Bonaparte, C.L. 1850. *Conspectus Generum Avium*, vol. 1. E.J. Brill, Lugduni Batavorum (Leiden).

Bredero, J.T., Heemskerk, W. & Toxopeus, H. 1977. Agriculture and livestock production in São Tomé and Príncipe (West Africa). Foundation for Agricultural Plant Breeding, Wageningen.

Brown, L.H., Urban, E.K. & Newman, K. 1982. *The Birds of Africa*, vol. 1. Academic Press, London.

Bruce, M.D. & Dowsett, R.J. 2004. The correct name of the Afrotropical mainland subspecies of Barn Owl *Tyto alba* Bull. Br. Orn. Club 124: 184-187.

Burlison, J.P. & Jones, P.J. 1988. *A report on the forestry resources and their present utilisation in the Democratic Republic of São Tomé and Príncipe.* International Council for Bird Preservation, Cambridge.

Cadée, G.C. 1981. Seabird observations between Rotterdam and the equatorial Atlantic. *Ardea* 69: 211–216.

Chapin, J.P. 1923. The Olive Ibis of Du Bus and its representative on São Thomé. *Amer. Mus. Novit.* 84: 1–9.

Chevalier, A. 1939. La végétation d l'île de San-Thomé. *Boletim da Sociedade Broteriana, 2ᵉ Ser.*, 13: 101–116.

Christy, P. 1990. New records of palaearctic migrants in Gabon. *Malimbus* 11: 117–122.

Christy, P. 1995a. Les pigeons "mangeurs de fumée". *Canopée* 5: 10.

Christy, P. 1995b. Ilhas Tinhosas – a unique and endangered seabird colony. *Gulf of Guinea Conservation Newsletter* 2: 2–3.

Christy, P. & Clarke, W. V. 1998. *Guide des oiseaux de São Tomé e Príncipe.* ECOFAC, São Tomé.

Christy, P. & Gascoigne, A. 1996. Príncipe Thrush rediscovered after more than 50 years. *Gulf of Guinea Conservation Newsletter* 4: 2–3.

CIA. 2001. *World Factbook.* CIA website http://www.cia.gov/cia/publications/factbook/

Collar, N.J. & Stuart, S.N. 1985. *Threatened Birds of Africa and Related Islands.* International Council for Bird Preservation (ICBP)/International Union for the Conservation of Nature (IUCN), Cambridge.

Collar, N.J. & Stuart, S.N. 1988. *Key Forests for Threatened Birds in Africa.* Monogr. 3, International Council for Bird Preservation, Cambridge.

Collar, N.J., Crosby, M.J. & Stattersfield, A.J. 1994. *Birds to Watch 2.* BirdLife International, Cambridge.

Correia, J.G. 1928-29a. Unpubl. bird lists and letters in AMNH.

Correia, J.G. 1928-29b. *The abits of ceveral birds of San Thome and Principe Is.* [*sic*] Unpubl. typescript in American Museum of Natural History.

Dallimer, M., King, T. & Leitão, P. 2003. New records of the São Tomé Grosbeak *Neospiza concolor.* Bull. Afr. Bird Club 10: 23-25.

Daudin, F.M. 1800. *Traité élémentaire et complet d'Ornithologie, ou Histoire naturelle des Oiseaux,* vol. 2. Buisson, Duprat, Deroy, Fuchs, Treuttel & Wurtz, and Villier, Paris.

Dohrn, H. 1866. Synopsis of the birds of Ilha do Príncipe, with some remarks on their habits and descriptions of new species. *Proc. Zool. Soc. Lond.* 324-327.

Dowsett, R.J. 1993. Afrotropical avifaunas: annotated country checklists. *Tauraco Res. Rep.* 5: 1-322.

Dowsett, R.J. & Dowsett-Lemaire, F. 1993. Comments on the taxonomy of some Afrotropical bird species. *Tauraco Res. Rep.* 5: 323-389.

Dowsett, R.J. & Forbes-Watson, A.D. 1993. *Checklist of Birds of the Afrotropical and Malagasy Regions.* Vol. 1: *Species limits and distribution.* Tauraco Press, Liège.

Drewes, R.C. & Wilkinson, J.A. 2004. The California Academy of Sciences Gulf of Guinea Expedition (2001). 1. The taxonomic status of the genus *Nesionixalus* Perret, 1976 (Anura: Hyperoliidae) treefrogs of São Tomé and Príncipe, with comments on the genus *Hyperolius.* Proc. California Acad. Sci. 55: 395-407.

Dutton, J. 1994. Introduced mammals in São Tomé and Príncipe: possible threats to biodiversity. *Biodiversity & Conservation* 3: 927-938.

Dutton, J. & Haft, J. 1996. Distribution, ecology and status of an endemic shrew, *Crocidura thomensis,* from São Tomé. *Oryx* 30: 195-201.

Eccles, S.D. 1988. The birds of São Tomé – record of a visit, April 1987 with notes on the rediscovery of Bocage's Longbill. *Malimbus* 10: 207-217.

Eck, S. 1995. Ergänzendes zu Morphologie und Systematik der *Speirops*-Arten (Zosteropidae). *Mitt. Zool. Mus. Berl.* 71 Suppl. Ann. Orn. 19: 101-107.

Eisentraut, M. 1965. Rassenbildung bei Saugetieren und Vogeln auf der Insel Fernando Poo. *Zoologischer Anzeiger* 174: 37-53.

Eisentraut, M. 1973. Die Wirbeltierfauna von Fernando Poo und Westkamerun. *Bonner Zool. Monogr.* 3: 1-428.

Erard, C. & Colston, P. 1988. *Batis minima* (Verreaux) new for Cameroon. *Bull. Br. Orn. Club* 108: 182-184.

Exell, A.W. 1944. *Catalogue of the Vascular Plants of S. Tomé (with Príncipe and Annobón).* British Museum (Natural History), London.

Exell, A.W. 1956. *Supplement to the Catalogue of the Vascular Plants of S. Tomé (with Príncipe and Annobón).* British Museum (Natural History), London.

Exell, A.W. 1963. Angiosperms of the Cambridge Annobón Island expedition. *Bull. Brit. Mus. (Nat.Hist.) Bot.* 3 (3): 93-118.

Exell, A.W. 1973. Angiosperms of the islands of the Gulf of Guinea (Fernando Po, Príncipe, S. Tomé, and Annobón). *Bull. Brit. Mus. (Nat.Hist.) Bot.* 4(8): 327-411.

Fa, J.E. 1991. *Conservación de los ecosistemas forestales de Guinea Ecuatorial.* International Union for the Conservation of Nature (IUCN), Gland.

Fa, J.E. & Juste, J. (eds). 1994. Biodiversity conservation in the Gulf of Guinea islands. *Biodiversity & Conservation* 3: 757-979.

Fahr, J. (1993). Ein Beitrag zur Biologie der Amphibien der Insel São Tomé (Golf von Guinea). *Faun. Abh. Mus. Tierkd. Dresden* 19: 75-84.

FAO. No date. *FAO Artemis NOAA-AVHRR NDVI Image Bank Africa 1981-1991.* Food & Agriculture Organisation of the United Nations, Rome.

Feiler, A. & Nadler, T. 1992. Über evolutive Beziehungen zwischen den Brillenvögeln der Gattung *Speirops* (Aves, Zosteropidae) von West-Kamerun und den Inseln im Golf von Guinea. *Bonn. zool. Beitr.* 43: 423–432.

Figueiredo, E. 1994a. Little known endemics collected by J. Espírito Santo in S. Tomé. *Garcia de Orta, Sér. Bot.* 12: 121–124.

Figueiredo, E. 1994b. New records for the flora of S. Tomé and Príncipe. *Garcia de Orta, Sér. Bot.*12: 125–126.

Figueiredo, E. 1994c. Diversity and endemism of angiosperms in the Gulf of Guinea islands. *Biodiversity & Conservation* 3: 785–793.

Figueiredo, E. 1994d. Contribution towards a botanical literature for the islands of the Gulf of Guinea. *Fontqueria* 39: 1–8.

Figueiredo, E. 1995. Floresta e endemismo em São Tomé e Príncipe. *Comun. IICT, Sér. Cien. agrárias.*19: 43–49.

Figueiredo, E. 1998. The Pteridophytes of São Tomé and Príncipe (Gulf of Guinea). *Bull. Nat. Hist. Mus. London (Bot.)* 28(1): 41–66.

Finsch, F.H.O. 1868. *Die Papageien Monographisch Bearbeitet*, vol. 2. E.J. Brill, Leiden.

Finsch, O. & Hartlaub, G. 1870. *Baron Carl Claus von der Decken's Reisen in Ost-Afrika*, vol. 4: *Die Vögel Ost-Afrikas*. C.F. Winter, Leipzig.

Fisher, B. 2004. Unusual nests of São Tomé Weaver *Ploceus sanctithomae*. Bull. Afr. Bird Club 11: 142-143.

Fishpool, L.D.C. & Demey, R. 1991. The occurrence of both species of "Lesser Golden Plover" and of nearctic scolopacids in Côte d'Ivoire. *Malimbus* 13: 3–10.

Fishpool, L.D.C. & Evans, M.I. 2001. *Important Bird Areas in Africa and Associated Islands: Priority sites for conservation.* Pisces Publications, Newbury and BirdLife International, Cambridge.

Fitton, J.G. 1987. The Cameroon line, West Africa: a comparison between oceanic and continental alkaline volcanism. In Fitton, J.G. & Upton, B.G.J. (eds) *Alkaline Igneous Rocks*. Geological Society Special Publication No. 30: 273–291.

Fitton, J.G. & Dunlop, H.M. 1985. The Cameroon line, West Africa, and its bearing on the origin of oceanic and continental alkali basalt. *Earth Planet. Sci. Lett.* 72: 23–38.

Fitton, J.G. & Hughes, D.J. 1977. Petrochemistry of the volcanic rocks of the island of Príncipe, Gulf of Guinea. *Contrib. Mineral. Petrol.* 64: 257–272.

Frade, F. 1958. Aves e mamíferos das Ilhas de São Tomé e do Príncipe – Notas de sistemática e de protecção à fauna. *Conf. Intern. Africanistas Ocidentais 6 Sessão Lisboa*, vol. 4: 137–149.

Frade, F. 1959. New records of non-resident birds, and notes on some resident ones, in São Tomé and Príncipe islands. *Proc. 1 Pan-Afr. Orn. Congr. (1957) Ostrich* Suppl. 3: 317–320.

Frade, F. & Naurois, R. de 1964. Une nouvelle sous-espèce de tisserin: *Ploceus velatus peixotoi* (île de São Tomé, golfe de Guinée). *Garcia de Orta* 12: 621–626.

Frade, F. & Vieira dos Santos, J. 1977. Aves de São Tomé e Príncipe (colecção do Centro de Zoologia). *Garcia de Orta, Sér. Zool.* 6: 3–18.

Fraser [L.] 1843a. Untitled [eight new species of Birds from Western Africa]. *Ann. Mag. nat. Hist.* 12: 478–479.

Fraser [L.] 1843b. Untitled [two new species of Birds from Western Africa]. *Proc. Zool. Soc. Lond.* 11: 34–35.

Fry, C.H. 1961. Notes on the birds of Annobón and other islands in the Gulf of Guinea. *Ibis* 103a: 267–276.

Fry, C.H. & Naurois, R. de 1985. *Corythornis* systematics and character release in the Gulf of Guinea islands. *Proc. 5 Pan-Afr. Orn. Congr.*: 47–61.

Fry, C.H., Keith, S. & Urban, E.K. 1985. Evolutionary expositions from 'The Birds of Africa': *Halcyon* song phylogeny; cuckoo host partitioning; systematics of *Aplopelia* and *Bostrychia Proc. Int. Symp. Afr. Vert.* (Bonn 1984): 163–180.

Fry, C.H., Keith, S. & Urban, E.K. 1988. *The Birds of Africa*, vol. 3. Academic Press, London.

Fry, C.H., Keith, S. & Urban, E.K. 2000. *The Birds of Africa*, vol. 6. Academic Press, London.

Gascoigne, A. 1993. A bibliography of the fauna of the islands of São Tomé e Príncipe and the island of Annobón (Gulf of Guinea). *Arquipélago Ciênc. biol. marin.* 11A: 91–105.

Gascoigne, A. 1995. *A Lista Vermelha de Animais Ameaçados de São Tomé e Príncipe.* ECOFAC, São Tomé.

Gascoigne, A. 1996. Additions to a bibliography of the fauna of São Tomé e Príncipe and the island of Annobón, Gulf of Guinea. Addendum *Arquipélago Ciênc. biol. marin* 14A: 95–105.

Giebel, C.G.A. 1872. Thesaurus Ornithologiae, vol. 1. Leipzig.

GGCG. 1996a. More records for the Dwarf Olive Ibis and São Tomé Fiscal Shrike. *Gulf of Guinea Conservation Newsletter* 3: 6.

GGCG. 1996b. New conservation project in Bioko. *Gulf of Guinea Conservation Newsletter* 4: 5.

GGCG. 1996c. Príncipe offshore Free Zone. *Gulf of Guinea Conservation Newsletter* 5: 2–3.

GGCG. 1996d. ECOFAC Phase 1: actions and achievements. *Gulf of Guinea Conservation Newsletter* 5: 5–6.

GGCG. 1997. Environmental assessments of Príncipe Free Zone project. *Gulf of Guinea Conservation Newsletter* 7: 3.

Gmelin, J.F. 1789. Systema Naturae, vol. 1, part 2. Lipsiae.

Goodwin, D. 1967. *Pigeons and Doves of the World* British Museum (Natural History), London.

Gray, G.R. 1849. *The Genera of Birds*, vol. 2. Longman, Brown, Green and Longmans, London.

Gray, G.R. 1862. Descriptions of a few West-African Birds. *Ann. Mag. nat. Hist.* (3)10: 443–445.

***Greeff, R.** 1884. Über die Fauna der Guinea-Inseln S. Thomé und Rolas. *Sitz-Ber. Ges. Beförd. ges. Naturw., Marburg* 2: 41–79.

Grimes, L.G. 1987. *The Birds of Ghana*. Check-list 9, British Ornithologists' Union, London.

Günther, R. & Feiler, A. 1985. Die Vögel der Insel São Tomé. *Mitt. Zool. Mus. Berlin* 61 *Suppl. Ann. Orn.* 9: 3–28.

Haffer, J. 1992. The history of species concepts and species limits in ornithology. *Bull. Br. Orn. Club* 112A: 107–158.

Haft, J. 1993. Ein Beitrag zur Biologie der Echsen der Insel São Tomé (Golf von Guinea), mit näherer Betrachtung zur Systematik von *Leptosiaphos africana* (Gray). *Faun. Abh. Mus. Tierkd. Dresden* 19: 59–70.

Hall, B.P. & Moreau, R.E. 1970. *An Atlas of Speciation in African Passerine Birds*. British Museum (Natural History), London.

***Hamilton, T.H. & Armstrong, N.E.** 1965. Environmental determination of insular variation in bird species abundance in the Gulf of Guinea. *Nature* 207: 148–151.

Harris, M.P. 1969. The biology of Storm Petrels in the Galapagos islands. *Proc. California Acad. Sci.* 37: 95–166.

Harrison, M.J.S. 1990. A recent survey of the birds of Pagalu (Annobón). *Malimbus* 11: 135–143.

Harrison, M.J.S. & Steele, P. 1989. *ICBP/EEC Forest Conservation Mission to São Tomé and Príncipe, January–March 1989. Report on Conservation Education and Training.* International Council for Bird Preservation, Cambridge.

Hartert, E. 1900. Untitled [*Chaetura thomensis*, sp. n.]. *Bull. Br. Orn. Club* 10: 53–54.

Hartert, E. 1928. A rush through Tunisia, Algeria and Marocco, and collecting in the Maroccan Atlas, in 1927. *Novit. Zool.* 34: 337–371. [Includes description of *Apus affinis bannermani*]

Hartlaub, G. 1848. Description de cinq nouvelles espèces d'oiseaux de l'Afrique occidentale. *Rev. Zool.* 108–110.

Hartlaub, G. 1849. Description de cinq nouvelles espèces d'Oiseaux de l'Afrique occidentale. *Rev. Mag. Zool.* 494–497.

Hartlaub, G. 1850. *Beitrag zur Ornithologie Westafrica's.* Verz. Öffentl. Privat-Vorlesung, Hamburg. [Reprinted in 1852 in *Abh. Geb. Naturw. Hamburg* 2(2): 1–56.]

Hartlaub, G. 1852. Zweiter Beitrag zur Ornithologie Westafrica's. *Abh. Geb. Naturw. Hamburg* 2(2): 57–68.

Hartlaub, G. 1853–54. Versuch einer synoptischen Ornithologie Westafrica's. *J. Orn.* 1: 385–400; 2: 1–32, 97–128, 193–218, 289–308.

Hartlaub, G. 1855. Bericht über die Leistungen in der Naturgeschichte der Vögel während des Jahres 1855. *Arch. Naturgesch.* 22: 1–32.

Hartlaub, G. 1857. *System der Ornithologie Westafrica's.* Shünemann, Bremen.

Hartlaub, G. 1861. Berichtungen und Zusätze zu meinem "System der Ornithologie Westafrica's". *J. Orn.* 9: 97–112, 161–176, 257–276.

Hartlaub, G. 1866. [Description of three new taxa from Príncipe, in Dohrn 1866.]

Heim de Balsac, H.H. & Hutterer, R. 1982. Les Soricidae (Mammifères, Insectivores) des îles de Golfe de Guinée: faits nouveaux et problèmes géographiques. *Bonn. zool. Beitr.* 33: 133–150.

Heine, F. & Reichenow, A. 1882–90. *Nomenclator Musei Heineani Ornithologici.* Friedländer & Sohn, Berlin.

Henriques, J. 1917. A Ilha de S. Tomé sob o ponto da vista historico-naturral e agricola. *Bol. Soc. Brot.* 27.

Herremans, M. 1988. Inter-island bird vocalisations on the Comoros. *Proc. 6 Pan-Afr. Orn. Congr.*: 281–295.

IGN. 1986. *Mapa de la República de Guinea Ecuatorial.* Madrid: Instituto Geografico Nacional.

Interforest. 1990. *Democratic Republic of São Tomé and Príncipe, 1: Results of National Forest Inventory, study of supply and demand of primary forest products – conclusions and recommendations.* Interforest AB, São Tomé.

IUCN. 1993. *Environmental Synopsis: São Tomé and Príncipe.* International Union for the Conservation of Nature (IUCN), Gland.

Jones, P.J. 1994. Biodiversity in the Gulf of Guinea: an overview. *Biodiversity & Conservation* 3: 772-784.

Jones, P.J. & Tye, A. 1988. *A Survey of the Avifauna of São Tomé and Príncipe.* ICBP Study Report 24 (64pp), International Council for Bird Preservation, Cambridge.

Jones, P.J., Burlison, J. & Tye, A. 1991. *Conservação dos Ecossistemas Florestais na República Democrática de São Tomé e Príncipe* International Union for the Conservation of Nature (IUCN), Gland.

Jones, P.J., Burlison, J. & Tye, A. 1992. The status of endemic birds and their habitats in São Tomé and Príncipe. *Proc. 7 Pan-Afr. Orn. Congr. (1988)*: 453–459.

Juste, J.B. 1996. Trade in the gray parrot *Psittacus erithacus* on the island of Príncipe (São Tomé and Príncipe, central Africa): initial assessment of the activity and its impact. *Biol. Conserv.* 76: 101–104.

Juste, J. & Ibañez, C. 1994. Bats of the Gulf of Guinea islands: faunal composition and origins. *Biodiversity & Conservation* 3: 837–850.

Juste, J., Ibañez, C. & Machordom, A. 2000. Morphological and allozyme variation of Eidolon helvum (Mammalia: Megachiroptera) in the islands of the Gulf of Guinea. *Biol. J. Linn. Soc.* 71: 359-378.

Keith, S. & Urban, E.K. 1992. A summary of present knowledge of the status of thrushes in the *Turdus olivaceus* species complex. *Proc. 7 Pan-Afr. Orn. Congr.*: 249–260.

Keulemans, J.G. 1866. Opmerkingen over de vogels van de Kaap-verdische Eilanden en van Prins-Eiland (Ilha do Príncipe) in de Bogt van Guinea gelegen. *Nederl. Tijdschr. Dierk.* 3: 363–401.

Keulemans, J.G. 1907. The Bronze Cuckoo (*Chrysococcyx smaragdineus*). *Bird Notes J. Foreign Bird Club* 5: 245–247.

Lack, D. 1958. The significance of the colour of turdine eggs. *Ibis* 100: 145–166.

Lains e Silva, H. 1958. São Tomé e Príncipe e a cultura do café. *Mem. Junta Invest. Ultramar* 1. Lisbon.

Lambeck, K. & Chappell, J. 2001. Sea level change through the last glacial cycle. *Science* 292: 679–686.

Lambert, K. 1988. Nächtliche Zugaktivität von Seevögeln im Golf von Guinea. *Beitr. Vogelkunde* 34: 29–35.

Lawson, W.J. 1984. The West African mainland forest-dwelling population of *Batis*; a new species. *Bull. Br. Orn. Club* 104: 144–146.

Lawson, W.J. 1986. Speciation in the forest-dwelling populations of the avian genus *Batis. Durban Museum Novitates* 13(21): 285–304.

Lee, D.-C., Halliday, A.N., Fitton, J.G. & Poli, G. 1994. Isotopic variations with distance and time in the volcanic islands of the Cameroon line: evidence for a mantle plume origin. *Earth Planet. Sci. Lett.* 123: 119–138.

Liberato, M.C. & Espírito Santo, J. 1972–1982. *Flora de São Tome e Príncipe.* Jardim e Museu Agrícola do Ultramar, Lisbon.

Loboch, M.Z. 1962. *Noticia de Annobón (Su geografía, historia y costumbres)* Papeleria Madrileña Mayor, Madrid.

Louette, M. 1981. The birds of Cameroon. An annotated check-list. *Verhandel. Kon. Acad. Wetensch. Lett. Schone Kunst Belg. Kl. Wetensch.* 43 (163): 1–295.

Loumont, C. 1992. Les amphibiens de São Tomé et Príncipe; revisions systematiques, cris nuptiaux et caryotypes. *Alytes* 10: 37–62.

Lowe, W.P. 1932. *The Trail That Is Always New.* Gurney & Jackson, London.

Mallett, J. 1995. A species definition for the Modern Synthesis. *Trends Ecol. Evol.* 10: 294–299.

Marshall, J.T. 1978. Systematics of smaller Asian night birds based on voice. *Orn. Monogr.* 25.

Mead, C.J. & Clark, J.A. 1987. Report on bird-ringing for 1986. *Ringing & Migration* 8: 135–200.

Melo, M.P. 1998. *Differentiation between Príncipe Island and mainland populations of the African Grey Parrot Psittacus erithacus: genetic and behavioural evidence and implications for its conservation.* First field expedition report 28 October–3 December 1998. Percy FitzPatrick Institute of African Ornithology, University of Cape Town, Cape Town.

Mey, E. 1993. *Cuculoecus africanus* n. sp. – ein neuer ischnozerer Kuckucksfederling von der Insel São Tomé (Golf von Guinea) (Insecta: Phthiraptera). *Faun. Abhandl. Staatl. Mus. Tierk. Dresden* 19: 103–110.

Mildbraed, J. 1922. *Wissenschaftliche Ergebnisse der zweiten Deutschen Zentral-Afrika-Expedition 1910–1911 unter Führung Adolf Friedrichs, Herzogs zu Mecklenburg. Band II: Botanik.* Klinkhardt & Biermann, Leipzig.

Ministério do Ultramar. 1958. *Carta de S. Tomé* Folhas 1-5. Ministério do Ultramar, Lisbon.

Ministério do Ultramar. 1962. *Carta de Príncipe* Folhas 1-2. Ministério do Ultramar, Lisbon.

Mitchell-Thomé, R.C. 1970. *Geology of the South Atlantic Islands.* Gebrüder Borntraeger, Berlin

Monod, T. 1960. Notes botaniques sur les îles de São Tomé et de Príncipe. *Bull. Inst. fr. Afr. noire* 22A: 19–83.

Monod, T., Teixeira da Mota, A. & Mauny, R. 1951. *Description de la Côte Occidentale d'Afrique (Sénégal au Cap de Monte, Archipels) par Valentim Fernandes (1506–1510).* Centro de Estudos da Guiné Portuguesa, 11, Bissau.

Monteiro, L.R., Covas, R., Melo, M.P., Monteiro, P.R., Jesus, P., Pina, N., Sacramento, A & Vera Cruz, J. 1997. Seabirds of São Tomé e Príncipe. Unpubl. rep., University of the Azores, Horta and BirdLife International, Cambridge.

Moreau, R.E. 1957. Variation in the western Zosteropidae. *Bull. Brit. Mus. Nat. Hist. Zool.* 4(7): 309–433.

Moreau, R.E. 1960. Conspectus and classification of the ploceine weaver-birds. *Ibis* 102: 298–321, 443–471.

Moreau, R.E. 1966. *The Bird Faunas of Africa and its Islands.* Academic Press, London.

Moreau, R.E. & Chapin, J.P. 1951. The African Emerald Cuckoo, *Chrysococcyx cupreus.* *Auk* 68: 174–189.

Nadler, T. 1993. Beiträge zur Avifauna der Insel São Tomé (Golf von Guinea). *Faun. Abhandl. Staatl. Mus. Tierk. Dresden* 19: 37–58.

Naurois, R. de 1972a. Avifaune terrestre et de rivages. *Bol. Brig. Fom. Agro-pec. São Tomé Príncipe* 6(23): 29–42.

Naurois, R. de 1972b. Noms portugais et noms indigènes des oiseaux de São Tomé et Príncipe. *Bol. Brig. Fom. Agro-pec. São Tomé Príncipe* 6(23): 43–46.

Naurois, R. de 1973a. L'avifaune marine des îles de São Tomé et de Príncipe. *Bol. Brig. Fom. Agro-pec. São Tomé Príncipe* 7(27): 33–43.

Naurois, R. de 1973b. Les ibis des îles de S. Tomé et du Prince: leur place dans le groupe des *Bostrychia* (=*Lampribis*). *Arq. Mus. Bocage* (2)4: 157–173.

Naurois, R. de 1975a. Le "Scops" de l'Ile de São Tomé *Otus hartlaubi* (Giebel). *Bonn. zool. Beitr.* 26: 319–355.

Naurois, R. de 1975b. Les Carduelinae des îles de São Tomé et Príncipe (Golfe de Guinée). *Ardeola* 21 especial: 903–931.

Naurois, R. de 1979. The Emerald Cuckoo of São Tomé and Príncipe islands (Gulf of Guinea). *Ostrich* 50: 88–93.

Naurois, R. de 1980. Le statut de *Halcyon malimbicus dryas* Hartlaub (Ile du Prince, Golfe de Guinée). *Bull. Inst. fond. Afr. noire* 42A: 608–618.

Naurois, R. de 1981. Les Phasianidae de l'île de São Tomé. *Cyanopica* 2: 29–36.

Naurois, R. de 1982. Une énigme ornithologique: *Amaurocichla bocagei* Sharpe, 1892. *Bull. Inst. fond. Afr. noire* 44A: 200–212.

Naurois, R. de 1983a. Les oiseaux reproducteurs des îles de São Tomé et Príncipe: Liste systématique commentée et indications zoogéographiques. *Bonn. zool. Beitr.* 34: 129–148.

Naurois, R. de 1983b. Falconidae, Psittacidae et Strigiformes des îles de São Tomé et Príncipe (Golfe de Guinée). *Bonn. zool. Beitr.* 34: 429–451.

Naurois, R. de 1984a. La moucherolle endémique de l'île de São Tomé, *Terpsiphone atrochalybeia* (Thomson 1842). *Alauda* 52: 31–44.

Naurois, R. de 1984b. *Prinia molleri* Bocage, 1887, endémique de l'île de São Tomé. *Riv. Ital. Orn.* 54: 191–206.

Naurois, R. de 1984c. Les *Turdus* des îles de São Tomé et Príncipe: *T. o. olivaceofuscus* (Hartlaub) et *T. olivaceofuscus xanthorhynchus* Salvadori (Aves Turdinae). *Rev. Zool. afr.* 98: 403–423.

Naurois, R. de 1984d. Le loriot endémique de l'île de São Tomé (Golfe de Guinée) *Oriolus crassirostris* (Hartlaub). *Cyanopica* 3: 121–134.

Naurois, R. de 1985. *Chaetura (Rhaphidura) thomensis* Hartert 1900 endémique des îles de São Tomé et Príncipe (Golfe de Guinée). *Alauda* 53: 209–222.

Naurois, R. de 1987a. Les Rallidae des îles de São Tomé et du Prince (Golfe de Guinée). *Cyanopica* 4: 5–26.

Naurois, R. de 1987b. Phalacrocoracidae et Ardeidae dans les Iles de São Tomé et du Prince (Golfe de Guinée). *Cyanopica* 4: 27–54.

Naurois, R. de 1987c. Notes on *Dicrurus m. modestus* (Hartlaub) and remarks on the *modestus* and *adsimilis* groups of drongos. *Bonn. zool. Beitr.* 38: 87–93.

Naurois, R. de 1988a. Les Columbidae des îles de S. Tomé et Príncipe. *Cyanopica* 4: 217–242.

Naurois, R. de 1988b. Note sur la pie-grièche *Lanius newtoni* (Bocage 1891), endémique de l'île de São Tomé (Golfe de Guinée). *Cyanopica* 4: 251–259.

Naurois, R. de 1988c. *Neospiza concolor* (Bocage, 1888) endémique de l'île de São Tomé (Golfe de Guinée). *Bull. Mus. Reg. Sci. Nat. Torino* 6: 321–339.

Naurois, R. de 1994. *Les Oiseaux des Iles du Golfe de Guinée/As Aves das Ilhas do Golfo da Guiné.* Instituto de Investigação Científica Tropical, Lisbon.

Naurois, R. de & Castro Antunes L.J. 1973. Repartição geográfica das espécies ornitológicas endémicas de São Tomé e Príncipe. *Bol. Brig. Fom. Agro-pec. São Tomé Príncipe* 7(27): 17–31.

Naurois, R. de & Wolters, H.E. 1975. The affinities of the São Tomé Weaver *Textor grandis* (Gray, 1844). *Bull. Br. Orn. Club* 95: 122–126.

Nill, T. 1993. Die Schlangen der Insel São Tomé (Golf von Guinea). *Faun. Abh. Mus. Tierkd. Dresden* 19: 71–73.

Nussbaum, R.A. & Pfrender, M.E. 1998. Revision of the African caecilian genus *Schistometopum* Parker (Amphibia: Gymnophiona: Caeciliidae). *Misc. Publ. Mus. Zool. Univ. Michigan* No.187: 1-32.

Peet, N.B. & Atkinson, P.W. 1994. The biodiversity and conservation of the birds of São Tomé and Príncipe. *Biodiversity & Conservation* 3: 851–867.

Pérez del Val, J. 1996. *Las Aves de Bioko.* Edilesa, León.

Pérez del Val, J. 2001. A survey of birds of Annobón Island, Equatorial Guinea: preliminary report. *Bull. Afr. Bird Club* 8: 54.

Peréz del Val, J., Castroviejo, J. & Purroy, F.J. 1997. Species rejected from and added to the avifauna of Bioko island (Equatorial Guinea). *Malimbus* 19: 19–31.

Pérez del Val, J., Fa, J.E., Castroviejo, J. & Purroy, F.J. 1994. Species richness and endemism of birds in Bioko. *Biodiversity & Conservation* 3: 868–892.

Peters, J.L. and others. 1931–1986. *Check-list of Birds of the World*, 15 vols. Harvard University Press and Museum of Comparative Zoology, Cambridge, Mass.

RDSTP. 1993. *Diário da República No. 13. Conselho de Ministros, Decreto-Lei No. 52/93* Government of São Tomé & Príncipe, São Tomé.

Reichenow, A. 1900–1905. *Die Vögel Afrikas.* J. Neumann, Neudamm.

Reinius, S. 1985. List of São Tomé and Príncipe birds at the Royal Museum of Natural History in Stockholm and a list of the birds of São Tomé and Príncipe. Unpubl. typescript on file at BirdLife International, Cambridge.

República de Guínea Ecuatorial. 1998. *Informe Interino del País a la Conferencia de las Partes del Convenio sobre la Diversificación Biológica.* Ministerio de Bosques y Medio Ambiente, Malabo.

Robins, C.R. 1966. Observations on the seabirds of Annobón and other parts of the Gulf of Guinea. *Stud. trop. Oceanogr. Miami* 4: 128–133.

Russell, G.J., Diamond, J.D., Pimm, S.L. & Reed, T.M. 1995. A century of turnover: community dynamics at three timescales. *J. Anim. Ecol.* 64: 628-641.

Salvadori, T. 1901. Due nuove specie di uccelli dell'Isola di S. Thomé e dell'Isola del Príncipe raccolte dal Sig. Leonardo Fea. *Boll. Mus. Zool. Anat. Comp. Univ. Torino* 16(414): 1–2.

Salvadori, T. 1902. On a new kingfisher of the genus *Corythornis. Ibis* (8)2: 566–569.

Salvadori, T. 1903a. Contribuzioni alla ornithologia delle isole del Golfo di Guinea. I. Uccelli dell'isola de Príncipe. *Mem. Reale Accad. Sci. Torino* (2)53: 1–16.

Salvadori, T. 1903b. Contribuzioni alla ornithologia delle isole del Golfo di Guinea. II. Uccelli di San Thomé. *Mem. Reale Accad. Sci. Torino* (2)53: 17–45.

Salvadori, T. 1903c. Contribuzioni alla ornithologia delle isole del Golfo di Guinea. III. Uccelli di Anno-Bom e di Fernando Po. *Mem. Reale Accad. Sci. Torino* (2)53: 93–125.

Sargeant, D.E. 1994. Recent ornithological observations from São Tomé and Príncipe islands. *Bull. Afr. Bird Club* 1: 96–102.

Sargeant, D.E. & Alexander-Marrack, P.D. 1990. São Tomé, 23–25 December 89. Unpubl. typescript.

Sargeant, D.E., Gullick, T., Turner, D.A. & Sinclair, J.C.I. 1992. The rediscovery of the São Tomé Grosbeak *Neospiza concolor* in south-western São Tomé. *Bird Conserv. Internat.* 2: 157–159.

Schätti, B. & Loumont, C. 1992. Ein Beitrag zur Herpetofauna von São Tomé (Golf von Guinea). *Zool. Abh. Staatl. Mus. Tierk. Dresden* 47: 23–36.

Schollaert, V. & Willem, G. 2001. A new site for Newton's Fiscal *Lanius newtoni*. *Bull. Afr. Bird Club* 8: 21-22.

Schultze, A. 1913. Die Insel Annobón im Golf von Guinea. *Petermanns Mittheilungen* 59: 131–133.

*****Sequeira, V.** 1994. Medicinal plants and conservation in São Tomé. *Biodiversity & Conservation* 3: 910–926.

Sharpe, R.B. 1869. On the birds of Angola – Part I. *Proc. Zool. Soc. Lond.* 563–571.

Sharpe, R.B. 1892a. Description of some new species of timeliine [*sic*] birds from West Africa. *Proc. Zool. Soc. Lond.* 227–228.

Sharpe, R.B. 1892b. *Catalogue of the Birds of the British Museum*, vol. 17. British Museum (Natural History), London.

Shelley, G.E. 1896–1912. *The Birds of Africa*. R.H. Porter, London.

Sibley, C.G. & Monroe, B.L. 1990. *Distribution and Taxonomy of Birds of the World*. Yale University Press, New Haven.

Snow, D.W. 1950. The birds of São Tomé and Príncipe in the Gulf of Guinea. *Ibis* 92: 579–589.

Sousa, J.A. de 1887. Aves da Ilha do Príncipe colligidas pelo Sr. Francisco Newton. *J. Sci. math. phys. nat. Lisboa* 12(45): 42–44.

Sousa, J.A. de 1888. Enumeração das aves conhecidas da Ilha de S. Thomé, seguida da lista das que existem d'esta Ilha no Museu de Lisboa. *J. Sci. math. phys. nat. Lisboa* 12(47): 151–159.

Stattersfield, A.J., Crosby, M.J., Long, A.J. & Wege, D.C. 1998. *Endemic Bird Areas of the World: Priorities for biodiversity conservation*. BirdLife International, Cambridge.

Stévart, T. & Oliveira, F. de 2000. Guide des Orchidées de São Tomé et Príncipe. ECOFAC-Gabon, Libreville.

Summers, R.W. & Waltner, M. 1979. Seasonal variation in the mass of waders in southern Africa, with special reference to migration. *Ostrich* 50: 21–37.

Summers, R.W., Cooper, J. & Pringle, J.S. 1977. Distribution and numbers of waders (Charadrii) in the southwestern Cape, South Africa, summer 1975–76. *Ostrich* 48: 85–97.

Themido, A.A. 1938. Aves das colónias portuguesas. (Catálogo das colecções do Museu Zoológico de Coimbra). *Mem. Estud. Mus. Zool. Univ. Coimbra* (1)110: 1–74.

Thomson, T.R.H. 1842. Description of a new species of *Genetta* and of two species of birds from western Africa. *Ann. Mag. Nat. Hist.* 10: 203–204.

Thys van den Audenaerde, D.F.E. 1967. The freshwater fishes of Fernando Poo. *Verhandel. Koninkl. Vlaamse Acad. Wetensch. Lett. Schone Kunst. Belg.* 100: 1–167.

Tutin, C.E.G. & White L.J.T. 1998. Primates, phenology and frugivory: present, past and future patterns in the Lopé reserve, Gabon. In D.M. Newbery, H.H.T. Prins & N. Brown (eds) *Dynamics of Tropical Communities*: 309–337 British Ecological Society Symposium No. 37. Blackwell Science, Oxford.

Tye, A. 1987. Identifying the major wintering grounds of palaearctic waders along the Atlantic coast of Africa, from marine charts. *Wader Study Group Bull.* 49: 20–27; 50: 17.

Tye, A. & Macaulay, L.R. 1993. The races of Olive Sunbird *Nectarinia olivacea* on the Gulf of Guinea islands. *Malimbus* 14: 65–66.

Tye, A. & Tye, H. 1987. The importance of Sierra Leone for wintering waders. In N.C. Davidson & M.W. Pienkowski (eds) The Conservation of International Flyway Populations of Waders: 71–75. *Wader Study Group Bull.* 49 Suppl./*IWRB Spec. Publ. 7.*

Urban, E.K., Fry, C.H. & Keith, S. 1986. *The Birds of Africa,* vol. 2. Academic Press, London.

Urban, E.K., Fry, C.H. & Keith, S. (eds). 1997. *The Birds of Africa,* vol. 5. Academic Press, London.

Verreaux, J. 1857. [Description of *Nectarinia hartlaubi,* in Hartlaub 1857.]

Verreaux, J. & Verreaux, E. 1851. Description d'espèces nouvelles d'oiseaux du Gabon (côte occidentale d'Afrique). *Rev. Mag. Zool. pure appl.* (2)3: 513–516.

V[ieira]., L. 1887. Aves da Ilha de S. Thomé. *Inst. Rev. Sci. Lit. Coimbra* (2)34: 562. [Introduces list written by Bocage (1887c): see Sousa 1888, p.154.]

White, F. 1983. Long-distance dispersal and the origins of the Afromontane flora. *Sonderbd. naturwiss. Ver. Hamburg* 7: 87–116.

White, F. 1983–84. Afromontane elements in the flora of S. Tomé: variation and taxonomy of some "nomads" and "transgressors". *Garcia de Orta, sér. Bot.,* 6(1/2): 187–202.

White, L.J.T. 1994. Patterns of fruitfall phenology in the Lopé Reserve, Gabon *J. Trop. Ecol.* 10: 289–312.

Wolters, H.E. 1983. Zur Systematik einiger Passeres aus Kamerun. *Bonn. Zool. Beitr.* 34: 279–291.

Zarske, A. 1993. Notizen zur Ichthyofauna der Insel São Tomé (Golf von Guinea) (Teleostei). *Faun. Abhandl. Staatl. Mus. Tierk. Dresden* 19: 85–88.

Maps and gazetteers consulted (not necessarily mentioned in the text)

Centro de Geográfia do Ultramar 1968. *Relação dos nomes geográficos de S. Tomé e Príncipe.* Junta de Investigações do Ultramar, Lisbon.

Centro de Informação e Turismo de S. Tomé e Príncipe. 1971. *Portugal. Província de S. Tomé e Príncipe. Edição Comemorativa do V Centenário do Descobrimento das Ilhas de S. Tomé e Príncipe* Centro de Informação e Turismo de S. Tomé e Príncipe.

Instituto Geográfico Nacional. 1986. *Mapa de la República de Guinea Ecuatorial (Annobón)* (1:40,000). Instituto Geográfico Nacional, Madrid.

Junta das Missões Geográficas e de Investigações Coloniais. 1948. *Atlas de Portugal Ultramarino.* Ministério das Colónias, Lisbon.

Ministério do Ultramar 1958. *Carta de S Tomé,* Folhas 1–5 (1:25,000). Ministério do Ultramar, Lisbon.

Ministério do Ultramar 1962. *Carta de Príncipe,* Folhas 1–2 (1:25,000). Ministério do Ultramar, Lisbon.

Office of Geography 1962. *Gazetteer No. 63: Rio Muni, Fernando Po, and São Tomé e Príncipe.* Dept of the Interior, Washington, D.C.

Oliveira, A.A. d' 1885. *Carta da Ilha de S. Thomé.* Commissão de Cartográphia, Lisbon.

Oliveira, A.A. d'(?). 1886. *Carta da Ilha do Príncipe* Commissão de Cartográphia, Lisbon.

Pontano, I.I. 1611. *Rerum et Urbis Amstelodamensium Historia.* ?Cornelius Haemodius?, Amsterdam. [Map between pp. 166 and 167.]

Servicio Geográfico del Ejército 1958. *Mapa Militar de la Guinea Española (1:40,000).* Madrid.

INDEX OF SCIENTIFIC NAMES

(page numbers in bold refer to the Systematic List)

INDEX OF ENGLISH NAMES

(page numbers in bold refer to the Systematic List)